A DEADLY WIND

A Deadly Wind

The 1962 Columbus Day Storm

JOHN DODGE

Oregon State University Press   Corvallis

Cover: The collapse of the Campbell Hall bell tower at the Oregon College of Education, now Western Oregon University, in Monmouth, Oregon, was the Columbus Day Storm image seen around the world. (*Photo by Wes Luchau*)

Share your storm stories and photographs and learn more about the Columbus Day Storm at the companion Facebook group: www.facebook.com/groups/columbusdaystorm

Library of Congress Cataloging-in-Publication Data
Names: Dodge, John (Columnist), author.
Title: A deadly wind : the 1962 Columbus Day storm / John Dodge.
Other titles: 1962 Columbus Day storm | Columbus Day storm
Description: Corvallis : Oregon State University Press, 2018. | Includes
    bibliographical references and index.
Identifiers: LCCN 2018016272 (print) | LCCN 2018018679 (ebook) | ISBN
    9780870719295 (e-book) | ISBN 9780870719288 (original trade pbk. : alk.
    paper)
Subjects: LCSH: Storms—Northwest, Pacific—History—20th century. |
    Storms—British Columbia—Pacific Coast—History—20th century. |
    Storms—California—Pacific Coast—History—20th century. | Storms—
    Pacific Coast (U.S.)—History—20th century. | Northwest, Pacific—History.
Classification: LCC QC943.5.U6 (ebook) | LCC QC943.5.U6 D63 2018 (print) |
    DDC 551.550979—dc230
LC record available at https://lccn.loc.gov/2018016272

♾This paper meets the requirements of ANSI/NISO Z39.48-1992 (Permanence of Paper).

First published in 2018 by Oregon State University Press
Third printing 2019
Printed in the United States of America

Oregon State University Press
121 The Valley Library
Corvallis OR 97331-4501
541-737-3166 • fax 541-737-3170
www.osupress.oregonstate.edu

Contents

To my parents, John R. Dodge and Corrine Baker Dodge

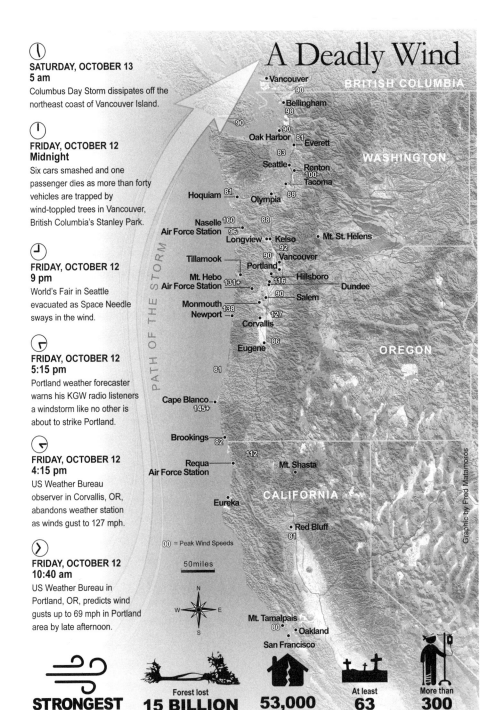

Preface

An unexpected phone call and a blog post ten days later from a well-known Pacific Northwest weather scientist became the two key ingredients that motivated me to write this book on the deadly 1962 Columbus Day Storm.

The phone conversation and blog post would not have mattered so much, if I hadn't experienced firsthand the strongest windstorm to strike the West Coast in recorded history. For me and hundreds of thousands of others, it was one of those seminal moments in life, on par with other events of the early 1960s never to be forgotten: astronaut John Glenn orbiting the Earth, the Cuban Missile Crisis, and the assassination of President John F. Kennedy.

I answered the call at my desk in the *Olympian* newsroom on October 5, 2012, a week before the fifty-year anniversary of the Columbus Day Storm. I had written in-depth feature stories about the storm on its twenty-fifth and fortieth anniversaries. Now a columnist for Washington State's capital city newspaper, I was looking for a fresh approach to my storm coverage. Bill Bruder, the caller on the other end, obliged.

Bruder's relationship to the storm was a unique one. In October 1962 he was serving as a US Navy weather observer at Naval Air Station Agana on the island of Guam. Against a backdrop of Cold War military tension that brought the world to the brink of a nuclear holocaust, Bruder and his crewmates took to the skies in Lockheed Constellation airplanes loaded with radar to look for Russian submarines and to track the fierce tropical storms that morphed into typhoons. The mission that kept him in the air the longest, and bedeviled him the most, was the tracking of Typhoon Freda in early October 1962.

Bruder, a rural resident of Mason County some thirty miles north of Olympia, talked to me in a matter-of-fact voice tempered by time and tinged with weariness. It was a condition caused by no one paying special attention to his story for too many years. After tracking Typhoon Freda for a week across thousands of miles of the vast Pacific Ocean, he had warned naval authorities in Pearl Harbor that the storm was likely to strike the West Coast and cause major damage. His warning did not register. His superiors told Bruder and his crew on October 9 to abandon the mission and fly back to Guam. The storm born from the heat of tropical ocean water would dissipate in the cooler water of the North Pacific Ocean, they said. These storms always do, they suggested. But that was a bit of an over-statement: the historical weather record shows that most of the damaging windstorms that strike the Pacific Northwest have a tropical storm component. Although no longer a typhoon when it reached the West Coast, Typhoon Freda provided much of the meteorological fuel for an unusually powerful nontropical cyclone.

I agreed to meet Bruder at an Olympia coffee shop to hear more about the life of a Cold War typhoon tracker and the tale of an aborted mission. I met a burly, goateed guy with swept-back hair and raptor-like eyes lit with the enthusiasm born of someone finally listening to his story. That first conversation with Bruder led to a front-page newspaper column in the *Olympian* headlined, "Shelton Man Did Job to Warn Navy about Deadly 1962 Storm." The first interview and initial account turned into the first step on the journey that became this book.

Then Cliff Mass, a University of Washington atmospheric sciences professor, offered his thoughts about the Columbus Day Storm three days after the storm's fiftieth anniversary. On his popular weather blog, which has had more than 36.6 million page views between its inception in 2008 through mid-March of 2018, he suggested that the storm would make a compelling topic for a book, even a movie. Mass studied under astronomer Dr. Carl Sagan at Cornell University, helping Sagan create a model of the Martian atmosphere. He also learned from his mentor the invaluable knack for deciphering the wonders of science in a way that captivates the general public.

Mass has made it his mission to bring the magic and power of weather and intense storms to life for those who read his weather blog and tune in to his weekly radio show, Fridays on KNKX-FM, a National Public Radio affiliate in Tacoma, Washington. The curly-haired professor with a rich, resonant voice ideal for radio spelled out a convincing case. "I have little doubt that if the Columbus Day Storm hit today, it would result in many billions, if not tens of billions of dollars of damage," he said. "The Columbus

Day Storm was far stronger than the Perfect Storm (Halloween 1991) of movie fame. I mean, it wasn't even close."

Mass called the Columbus Day Storm the strongest nontropical windstorm to strike the West Coast in recorded history and perhaps the mightiest nontropical cyclone winds to reach the lower forty-eight states. He placed it in the same realm as Superstorm Sandy, the extratropical cyclone that struck the East Coast on October 29, 2012, fifty years after the Columbus Day Storm. The measure of Superstorm Sandy outpaced the Columbus Day Storm on most fronts, including deaths and destruction, storm surge, and atmospheric pressure at the storm center. But the Columbus Day Storm winds were stronger. "You would think someone would have written a gripping book about the storm," Mass wrote in his October 15, 2012 blog. "But no one has. Unimaginable. What an opportunity for a good writer. This storm has everything." Then he pitched the storm with unbridled enthusiasm, listing some of the storm's eye-catching traits: winds exceeding 150 miles per hour and a storm equal to Hurricane Katrina, huge swaths of forests leveled by the winds, dozens of deaths, hundreds of injuries, and tens of thousands of homes and buildings destroyed or seriously damaged.

Mass wasn't done selling the storm, adding that it triggered record flooding and deadly landslides in California, widespread power outages throughout the Pacific Northwest, dramatic stories of rescues and recoveries, all with very little advance warning of the storm's might. "But valiant meteorologists get the word out that day, after offshore ships are hit by extraordinary winds, seas, and low pressure," he added. "What happened that day was so extreme, no exaggeration is needed."

The storm book sales pitch by Mass piqued my interest, rekindling memories of my own experience, gathered with friends and family in an Olympia-area home battered by the storm. I began some initial internet research to verify that Mass's assertion that no one had written a book about the Columbus Day Storm was correct. I concluded there were no full-length nonfiction books detailing the storm's development, the victims, and the storm's consequences. Two publications, however, are worth noting. Portland author and historian Ellis Lucia wrote *The Big Blow*, a fast-paced sixty-five-page summary of the storm, which was published in 1963 and well received by the public. At about the same time, Dorothy Franklin, who was a Salem, Oregon–area nurse and mother of five, collected enough newspaper article information and photographs to compile a spiral-bound, 180-page book printed by Gann Publishing of Portland, Oregon, called *West Coast Disaster.* The two early contributions of the Columbus Day Storm historical record can be purchased online or found in many of the region's libraries.

Mass was right: there was a lot of room for a journalist to venture forth on a path of her or his own. That's what I did. I conducted Columbus Day Storm research for about four months before I requested a mid-February 2013 meeting with Mass at his office on the University of Washington campus. My purpose was twofold: I wanted him to know I was working on a book about the storm. But more importantly, I was curious to find out whether any other authors or journalists had stepped up to the challenge.

His sixth-floor office in the Atmospheric Sciences-Geophysics Building seemed small for such an outsized personality: cabinets and shelves crammed with books and files, binders filled with research papers, and a desk with one extra chair. He greeted me with a certain wariness, not sure what to make of me. But he warmed up as I explained my personal connection to the storm, my journalistic background, my initial research, and my desire to write this book on my own terms. "Why are people so fascinated by severe weather and, in particular, the Columbus Day Storm?" I asked.

"It was a force of nature so much greater than man," Mass said in near-reverent terms. "This was almost like a religious event, not just a scientific event." If weather-watching was a religion in the Pacific Northwest, Mass would be the high priest, a stature he embraces. And, he said, the storm's sudden assault on the Pacific Northwest was so intense and unexpected, it magnified the storm's mystique. "Weather forecasting was so unbelievably primitive at the time," he said. He shared with me his file folder filled with maps, photographs, and reports about the Columbus Day Storm, material he compiled in writing his 2008 book about Pacific Northwest weather. I told him I had much, if not all, of the same material. That seemed to convince him I was serious.

He suggested I enroll in his Atmospheric Sciences 101 class to hone up on the weather science associated with the storm. I noted that I had taken the class as a UW student in 1966. Mass, who received his PhD in atmospheric sciences from the UW in 1978, asked me who taught the class. I didn't remember. Finally, I asked, "Have any other writers expressed a desire to write this book?" His answer was, "No." I breathed a sigh of relief and left our meeting more determined than ever to take ownership of the Columbus Day Storm story.

The research, the travel along the path of the storm, and the writing stretched out over the years. Most important to me were the hundreds of accounts of storm survivors living in Northern California, Western Oregon, Western Washington, or southern British Columbia on October 12, 1962. Whenever I mentioned my book project to someone who lived through the storm, they immediately launched into their personal narrative. Some stories were ordinary: a beloved cherry tree was blown over in the backyard or meals had to be cooked on a woodstove because the power was out for three days. Other memories were sensational, dramatic, tragic: a daughter still struggling with the loss of her parents, who were crushed to death in their car by a falling tree; a young boy, now a middle-aged man, mauled by an escaped lion; a husband and wife trapped on their kitchen floor while the winds tore apart the home they had just built.

The storm survivors I talked to often would ask me, "Why are you writing this book?" Now that I've done my best to answer that question, it's time to join me on this journey along the path of the 1962 Columbus Day Storm.

Acknowledgments

So many people played important roles in helping me write this book. It would be next to impossible to name then all, but it's my pleasure to mention some of the key contributors to *A Deadly Wind.*

I enlisted the help of several students to scour the microfilm of newspaper coverage of the 1962 Columbus Day Storm. They were Casey Wyrick and Mackenzie Emerson at Black Hills High School in Tumwater, Washington; Kristina Hernandez and Stacer McChesney at the University of Oregon in Eugene; and Jonny Wakefield at the University of British Columbia in Vancouver, British Columbia, Canada. They saved me time and spared me from eye fatigue and headaches.

In the early years of book research, I practically lived at the Washington State Library in Tumwater, Washington. I read dozens of books from the library's Northwest Collection, gaining insight about the communities along the path of the mighty windstorm and learning more Pacific Northwest history. State librarians Sean Lanksbury, Crystal Lentz, Kathryn Devine, Mary Schoff, Kim Smeenk, and Kathleen Roland were always generous with their expertise and time.

Other libraries and librarians offered research assistance, including Denise Reilly, adult services librarian, Newberg Public Library in Newberg, Oregon; Larry Landis and Rachel Lilley, Special Collections and Archives Research Center at the Oregon State University Valley Library in Corvallis; Tamara L. Vidos, University of Oregon Knight Library, Eugene; and the helpful staff in the Northwest Room of the Tacoma Public Library, Tacoma.

The region's historical societies and museums were rich in documents, personal storm accounts, and photographs that helped add depth and texture to the book. Among those who lent a hand were Scott Daniels, reference services manager at the Oregon Historical Society Research

Library, Portland; Sachiko Otsuki, museum collections specialist, Lincoln County Historical Society, Newport, Oregon; and Bill Watson, collections curator, Cowlitz County Historical Museum, Kelso, Washington. Kathryn Dysart gave me a guided tour of the Oregon State Hospital Museum of Mental Health in Salem. Also in Salem, Andrew Needham at Oregon State Archives shared with me storm-related documents pertaining to the Oregon State Hospital.

I received inspiration and guidance from several authors and journalists. Olympia author Jim Lynch was my literary compass on this book project, influencing me every step of the way. Olympia author Maria Ruth was an enthusiastic, steady companion, helping me navigate the bumpy road of nonfiction book writing, editing, and publishing. Craig Welch, a *National Geographic* writer and longtime environmental reporter for the *Seattle Times* showed me how to write a nonfiction book proposal, and Washington state historian John Hughes, my mentor and former editor and publisher of the *Daily World* in Aberdeen, Washington, displayed a fierce affinity for the book from start to finish.

Fellow journalists and Oregon State University Press authors Floyd McKay and R. Gregory Nokes read draft manuscripts of the book and were constructive with their praise and criticism of the project. Other faithful readers of early drafts and single chapters were my dear friends, Steve and Sandy Wall; my daughter, Kathryn Dodge; and my wife Barbara Digman, who was by my side, encouraging and comforting me on the years-long journey to publication.

My treatment of wind, weather forecasting, and meteorology was aided in large part by Wolf Read, a British Columbia–based forest climatologist who knows more about the history of windstorms in the Pacific Northwest than anyone else; Jim Holcomb, a meteorologist who had the daunting task of trying to forecast the weather the day the Columbus Day Storm struck the West Coast; Ted Buehner, a warning coordination meteorologist for the National Weather Service and, as a young boy, a Portland witness to the storm; and Cliff Mass, the University of Washington atmospheric sciences professor whose love of severe weather is unbridled and infectious.

I also want to thank my literary agent, George Lucas of Inkwell Management in New York, for helping me bring order to the book. The kind folks at Oregon State University Press, including Mary Elizabeth Braun, Micki Reaman, and Marty Brown, have provided a welcoming home for *A*

Deadly Wind. The deft touch of copy editor Susan Campbell saved me from any number of embarrassing errors. Thank you, Susan. *A Deadly Wind* also benefited from the keen eye and skills of indexer Mary Harper.

Finally, I owe a huge debt of gratitude to all the storm survivors who shared their experiences with me. You are the heart and soul of this book, and I am forever grateful.

1

Out on a Limb

Meteorologist John C. "Jack" Capell opened the door to a small radio broadcast room, really not much more than a closet, at the US Weather Bureau (precursor to the National Weather Service) office on Portland's Marine Drive. The well-known KGW television and radio weather fore-caster stepped up to the microphone to deliver his 5:15 radio weather forecast on October 12, 1962. He took a deep breath and issued a severe weather warning for the ages: "The worst storm I have ever seen is approaching Portland and the Willamette Valley right now," Capell told his radio audience in the clipped, staccato-style voice of the day. "I have never seen reports like this before. This is going to be a storm that causes a lot of damage." Thousands of KGW AM-620 radio listeners in northwest Oregon and southwest Washington heard Capell's bold message, delivered as the Friday evening rush-hour traffic in Portland began to build. Many may owe their lives to his last-minute warning of a damaging, lethal storm about to strike.

Capell knew the craft of weather forecasting was part emerging science and part an artful game. Bits and pieces of information, combined with intuition born of confidence, guided him. He relied on morning reports of an advancing windstorm received from storm-battered ships at sea, plunging barometric pressure readings, and a 4 p.m. report of winds gust-ing to 86 miles per hour in Eugene, one hundred miles south of Portland. Those hurricane-force winds were unprecedented in the Willamette Valley interior. Soon after the wind was reported, the Eugene station blacked out, followed by all the other secondary weather stations south of Portland. "That's the only way we could tell where the storm was," Capell said of the power outages marching north up the Willamette Valley. "There was no power. There were no reports." Next Portland experienced a sudden spike in air temperature, jumping from 55 degrees Fahrenheit at 4:40 p.m. to 64 degrees at 5:08 p.m.

It all combined into a forecast for the ages: a dangerous, ferocious windstorm was about to strike the Rose City. Capell was out on a limb all alone with his storm prediction. At 5 p.m., the ever-conservative weather bureau hadn't budged from its Friday morning forecast of winds gusting to 69 miles per hour by late afternoon in the Portland area. Capell, a former United States Weather Bureau meteorologist who joined the KGW news team full time in 1958 after working part time at the station since its inception in 1956, was not prone to sensational weather predictions. But the World War II veteran of D-Day and the Battle of the Bulge felt compelled to sound the alarm during the Friday night rush-hour commute.

The thirty-nine-year-old weather forecaster was no stranger to taking decisive action or working under duress. He was among the first D-Day troops to land on June 6, 1944, on Utah Beach in Normandy, France. He stayed on the beach for hours to repair a waterlogged army Jeep he was assigned to, German bullets and mortars flying all around him. He fought in the trenches and foxholes for months without hot food, a cot, or a shower. He had one day at a rest camp before he was called back to the front lines to fight the Battle of the Bulge and other bloody encounters with the German troops leading to VE Day on May 8, 1945.

Capell, a sturdy, athletic man with a thick head of black hair, somber, penetrating eyes and a smallish mouth set above a rounded chin, didn't often talk about his wartime experiences with friends or colleagues, including future Oregon governor Tom McCall, who worked in the small, tight-knit newsroom that was pioneering Pacific Northwest television news broadcasting at KGW, an NBC affiliate. Some confidants knew he was a private and a rifleman in the Fourth Infantry Division, which numbered fifteen thousand men at full strength. World War II casualties among the division's original and replacement troops reached a staggering thirty thousand, more than any of the other sixty-two army divisions assigned to the campaigns in Europe and North Africa.

Even fewer knew what motivated him to pursue a career in weather forecasting. The career choice was born during his bloody, grueling, eleven-month journey from the beaches of Normandy through France, Belgium, the deadly Hürtgen Forest of Germany, the Battle of the Bulge, and the liberation of Dachau concentration camp prisoners. Death and deprivation were Capell's constant companions as he moved from frontline foxholes through sniper-filled, bombed-out villages strewn with bodies of civilians and soldiers. Capell was wounded once when a German artillery shell hit near a can of gasoline, catching Capell on fire. In the final days of February, in the Siegfried Line trenches eleven miles east of St. Vith, Belgium, Capell used a Browning automatic rifle, which is a light machine gun, to shoot down a German two-engine fighter-bomber. All these experiences were mind-numbing and traumatic, but there was one brief moment in a foxhole near Montebourg, France, that forever changed the trajectory of Capell's life in a positive way.

The weather the week before and after D-Day had been dull and cloudy, with only an occasional sun break. Capell was sitting in a foxhole in mid-June when the dreary cloud cover peeled away, revealing bright blue skies and towering white clouds glistening in the warm sunshine. The shelling by the German troops seemed to momentarily subside. "I began to feel better immediately and filled with hope that I might survive," he wrote years later. "It was a brief moment of spiritual renewal for me as I gazed at the beauty that still could be seen in the world, and the feeling that not all was misery and horror."

And then the war-weary soldier had an epiphany: "I was so intrigued by the sharp outline of the gleaming white clouds that I wanted to learn more about them. It was at this moment that I determined that if I did survive I would someday study what made such beautiful formations in the sky," he said. "I could not have foreseen that this vision and these dark days at war would influence my later life and be a major factor of my career choice." After the war ended, Capell returned to Seattle and used the GI Bill to enroll at the University of Washington, where he majored in atmospheric sciences. He graduated in 1949, married Sylvia E. Wagner in 1951, and embarked on a lifelong career as a meteorologist.

After taping his Friday night weather report, Capell stepped outside the weather bureau's drab concrete building at the Portland International Airport and cast an anxious eye past the Portland skyline to the south and west in search of a mega-windstorm. He felt a touch of panic: the sun was setting in a clearing sky and the winds barely rustled. Driving back to his downtown television studio for the 6 p.m. newscast, Capell turned on the car radio and listened to his words. "I thought, 'Oh my god, did I really

Portland meteorologist Jack Capell delivers an early 1960s weather report from the KGW-TV studio. Courtesy of Tom Capell.

say that? The storm has missed us, or it's petered out. How am I going to explain this?' I was really scared."

Suddenly, the sky turned an eerie yellowish-green from light scattered through dense clouds some twenty miles to the south. As the clouds advanced the sky turned bruised and dark. The winds began to strengthen and shriek. Wind-blown roofing, glass, and street signs flew through the air past his car. The Columbus Day Storm of 1962 had reached the Rose City, advancing on a lethal path that began in Northern California in the morning and would push through the Puget Sound region of Washington State and southern British Columbia before finally losing strength in the early morning hours of October 13. It was a fierce, freakish windstorm, stronger than any other nontropical storm ever to strike the West Coast, and perhaps the lower forty-eight states. In a twisted way, Capell was a relieved and satisfied guy as chaos swirled around him.

Capell arrived at the KGW Broadcast House at 1139 SW Thirteenth Avenue in downtown Portland at about 5:40 p.m. He entered a newsroom lit by candlelight and filled with coworkers unsettled by the storm, anxious for an explanation from their broadcast meteorologist. In a calm, confident voice he told listeners that he had witnessed widespread damage and flying debris driving across town from the weather bureau office at the Portland airport. He said the strongest low-pressure center that he has seen since December 1951, measuring 28.85 inches, was centered just off the mouth of the Columbia River, and high pressure to the south was driving violent winds north up the Willamette Valley. Reports from the south were spotty because of widespread power outages.

He then turned his attention to the KGW wind gauge mounted on the roof. "There's 85 miles an hour—the highest gust I've ever seen," he said, his voice rising. "And there's another 85 on our gauge and, what's that, a 93—and there goes our gauge. The wind gauge just went." The winds also toppled the KGW television transmission tower on Skyline Boulevard. The 648-foot-tall steel and cable structure was built to withstand winds of 100 miles per hour. It was no match for this wind.

The downtown streets and sidewalks were filled with Friday rush-hour traffic and pedestrians. Many were unaware of the advancing storm. Priscilla Julien and a friend were caught downtown as the storm struck, walking along Washington Street about 5:45 p.m., trying to reach the friend's car parked under the Morrison Street Bridge. As they crossed SW Fourth Avenue, the wind blew Julien off her feet and sent her sailing and crashing to the pavement. She spent the next three weeks in the hospital, recovering from knee surgery and a chipped bone in the other knee. That wasn't all she suffered: a falling tree crushed her Plymouth sedan and the wind tore the chimney and a section of roofing off her Portland home. From her hospital bed she had to contact a roofer, tree surgeon, and chimney repair company. At that time all the problems had seemed insurmountable, but time had erased most of the angst and frustration, she said.

Mary McDonald Eden was a nineteen-year-old student at Phagan's Beauty School in downtown Portland, trying to get home to the Portland suburb of Milwaukie. Eden recalled what it was like, driving east across the Willamette River on the Steel Bridge, with its honeycomb mesh of steel gratings offering views of the roiled, feculent river water below. She crossed the bridge shortly before Capell did on his way back to the KGW station. The wind was blowing downstream and under the bridge, lifting the car off the bridge deck and slamming it back down again, she said. To the south, a semitruck was lying on its side on the Burnside Bridge. The driver, who had scrambled from the overturned truck, was hanging on to a light pole to keep from being blown into the river. After crossing the river and heading south on McLoughlin Boulevard, Eden watched the wind pick a camper off the back of a pickup truck and fling it into the air. Downed trees and power lines blocked three of the road's four lanes. Unable to navigate the storm mess, Eden holed up that night in the Milwaukie beauty salon where she worked part time.

"Our advice is to stay inside—stay away from windows at any cost for heaven's sake," KGW radio broadcaster Wes Lynch told listeners in a serious but not excitable voice. "There's some danger of getting hurt anywhere in Portland. If you're driving, find a place to park that car and get inside a public building." The downtown hotels—including the 1927 Heathman Hotel with its Italian Renaissance exterior and art deco interior, and the baroque revival–style Benson Hotel, built in 1913 and sheathed in red brick and cream-colored glazed terra cotta, the product of lumber baron and teetotalling philanthropist Simon Benson—were filling with football fans when the storm struck. The visitors to town gathered in hotel lobbies and downtown bars and restaurants, drawn to a Saturday afternoon West Coast collegiate clash at Portland's Multnomah Stadium between the Oregon State University Beavers, led by Heisman Trophy candidate and quarterback Terry Baker, and the formidable University of Washington Huskies from Seattle, Washington, a city basking in the glory of the continuing success of its World's Fair.

The out-of-towners were soon sharing the hotel lobbies, bars and rooms with panic-stricken Portlanders caught downtown, seeking shelter from flying shards of glass, street signs, and other windblown items, including men's fedora hats, still very much in style. "It's packed like a can of worms down here," Benson Hotel manager Joe Callihan said as the storm raged into the night. "We have people lined up in the lobby without rooms."

The winds were strong enough to crumple two 520-foot-tall Bonneville Power Administration main transmission towers carrying electricity across the Columbia River between Washington and Oregon. They were built to withstand winds of 100 miles per hour. Television and radio transmission towers in the Portland hills west of town were smacked to the ground, no match for the hurricane-force winds. It was 6:30 p.m. and the darkening city was a scene of chaos and destruction as debris from towering construction projects rained down on the streets below. All five Portland television stations were knocked off the air by the windstorm at various times Friday night. Most people had no idea what was happening on the weather front. Portlanders early Friday night were largely unaware that Oregon coastal cities and Willamette Valley towns to the south, the likes of Newport, Coos Bay, Eugene, Corvallis, and Salem, were already being dashed and torn apart by the strongest winds ever recorded on the West Coast, topping 100 miles per hour in several places. Widespread power outages to the south of Portland had isolated the stricken towns.

Capell was a pioneer television and radio weather forecaster who had earned the trust of his listening and viewing audience when the storm struck. If Capell said this would be a storm for the ages, his radio listeners believed him. But near blackout conditions limited his listening audience to those who had transistor radios that picked up a KGW radio signal, which was relying on auxiliary power produced by an emergency generator. Transistor radios, which had just penetrated the fledgling mass-market world of consumer electronics in 1958, were not yet a staple in everyone's homes. Capell and his colleagues stayed on the air into the early morning hours Saturday, relaying to their anxious radio audience bits and pieces of information gathered from phone calls coming in to the station. KGW was a lifeline on a night of weather terror never before experienced by the residents of the Pacific Northwest.

2
Tracking Typhoon Freda

Two weeks before weather forecaster Jack Capell sounded his storm alarm in Portland, Oregon, United States Navy weather observers stationed nearly 5,700 miles across the Pacific Ocean in Guam took note of a weather disturbance near Enewetak Atoll in the Marshall Islands. The developing storm's first breath some 2,400 miles southwest of Hawaiʻi was discernible, but nothing to cause alarm. Two parental breezes, perhaps a wispy warm trade wind called an easterly wave and a slightly cooler westerly flow from a passing thunderstorm, gave birth to what became Typhoon Freda, a key ingredient in the deadly Columbus Day Storm that struck the West Coast on October 12, 1962. The breezes converged 20 degrees west of the international date line and just far enough north of the equator—about 15 degrees latitude—to experience the force of the spinning earth known as the Coriolis effect, which sends winds moving in a counterclockwise direction in the Northern Hemisphere.

Hundreds of these tropical weather disturbances are of no consequence, short-lived and lacking the meteorological features they need to grow powerful. There's a touch of irony in the creation of a typhoon: A mature typhoon is a mighty display of destruction and power. But at birth, it is as vulnerable as a newborn child. The birth of a tropical storm requires conditions to be just right. The upper atmosphere must be stable. Wind shear, which is the change in wind speed with height, must be negligible to keep from poking holes in the big cumulus clouds that form in the early stages of storm development.

There's more to the tropical cyclone recipe: The air must be moist near the mid-level of the troposphere, which stretches to some seven miles above the earth and represents that part of the atmosphere where clouds and weather reside. And the atmosphere needs to cool fast enough with height to allow convection—the vertical movement of air. Finally and of

utmost importance, the ocean must be warm—at least 80 degrees Fahrenheit—and relatively shallow, not much more than 150 feet deep. The warm
water is the fuel source, the energy, the lethal ingredient of any tropical
storm.

On September 28, 1962, this fledgling storm was passing all the tests.
A typhoon tracking squadron based at Naval Air Station Agana on the
island of Guam was assigned to monitor the developing storm. Bill Bruder,
a twenty-year-old Texan, was a member of that squadron. Just four months
earlier he had completed his training as a navy weather observer at the
Lakehurst navy base, in New Jersey, and been assigned to a combined
operation established in 1959 among the US Weather Bureau, navy, and air
force. It was called the US Fleet Weather Central/Joint Typhoon Warning
Center, with headquarters at Pearl Harbor, Hawaiʻi. Against a backdrop of
Cold War military tension between the United States and the Soviet Union
that was about to reach new heights, Bruder and his crewmates took flight
to track the fiercest of tropical storms.

Bruder's tour of duty on Guam, the westernmost point of US-con-
trolled soil in the vast Pacific Ocean, lasted seventeen months and included
forty-four airborne missions, typhoon tracking assignments, and synoptic
flights to confirm the birth of low-pressure systems that had the potential
to grow into dangerous storms. The mission that kept him in the air the
longest, and haunted him the most through the years, was that early October flight in 1962. Bruder, an Austin, Texas, native, was born in October
1941, the son of a short-haul truck driver dad and homemaker mom. His
fascination with weather ran deep. As a young boy, he loved to watch the
heat lightning and thunderstorms sweep across the Austin sky, delivered
from the Gulf of Mexico in the spring and early summer months. One such
storm paid a visit to his modest Austin neighborhood in June of 1953. The
storm ratcheted his weather curiosity up a notch or two to a place it would
stay the rest of his life.

"I remember telling my mom I wanted to sleep on the garage roof
under the stars," Bruder recalled. The stars shined bright in a south Texas
sky unencumbered by light pollution from population growth, space satellites, or twenty-first-century commercial jet traffic. The night passed
without incident, dawn broke, and the morning started out uneventful as
well. "I woke up to the birds chirping. Suddenly, it got real quiet. I peeked
out from under my sleeping bag and saw the sky turn a yellowish-green. I

knew something was happening so I jumped up and climbed off the roof. Then 'Boom!' A tornado hit. I saw a garbage can and debris flying through the air and the porch lift off our neighbor's house. Fortunately it was a small tornado. It jumped our house and didn't cause much more damage."

That close encounter with a tornado piqued his interest in meteorology, an interest that simmered within Bruder through his teen years at A. N. McCallum High School. He joined the US Navy Reserve his senior year, and when he graduated in 1960, he enlisted in the navy. After boot camp, he reported in October 1961 to aerographer's mate school at the Lakehurst navy base in New Jersey. He spent the next six months immersed in meteorology and oceanography, fine-tuning his ability to observe and forecast weather. Upon graduation he was assigned to typhoon tracking duties in Guam. What better place to experience weather than the North Pacific Ocean, home to more destructive cyclones than anywhere else in the world.

The budding typhoon began to take shape as ascending whooshes of warm water vapor drawn from a vast ocean reservoir of warm tropical water began to rise, causing the atmospheric pressure at the surface of the ocean to drop. Like a honeybee attracted to a blossoming flower, denser surface air rushed in to fill the void. That air in motion equaled a 15-mile-per-hour wind, nothing out of the ordinary, but a precursor of things to come. And as the water in a gaseous state rose higher in the atmosphere, it cooled and condensed to form billowing cumulus clouds. Then a cavalcade of meteorological events both spontaneous and reactive began to unfold. The condensation of water vapor released latent heat that, in turn, caused more water vapor to rise, which triggered more condensation, more clouds and more thermal energy to reheat the air, driving it higher and higher. At the same time, the storm-making machinery thousands of feet above the ocean surface created even lower pressure below, which amped up those original breezes at what now was the center of a developing storm. Those winds kept spinning like a top, always counterclockwise north of the equator thanks to the Coriolis effect.

The atmosphere and the ocean, so different and yet so much the same, working together as one large organism to support life and cause havoc. In this case they were in a tempestuous mood, cooking up a stormy stew that could increase with intensity surely enough to secure a name. The birthplace of this rapidly developing tropical storm was in a cluster of islands known as Micronesia. More precisely, it was close to the Enewetak Atoll, an elliptical necklace of coral reef islands sandwiched between sandy white beaches and a tranquil turquoise lagoon, just a speck of low-lying, ancient geography at the northwest edge of the Marshall Islands.

From high aloft, the atoll connotes an image of peace and quietude, a slight interruption in a vast, azure sea. But look a little closer and the vestiges of a violent past begin to emerge. From 1948 to 1958, the United States displaced the inhabitants and turned the atoll into ground zero of the Pacific Proving Grounds: Enewetak was home to forty-three nuclear tests, including the first hydrogen bomb blast—code name Ivy Mike—in 1952. Nuclear bomb blast craters pockmark the forty islands covering some 2.26 square miles of land. Two islands and a part of a third were vaporized by nuclear explosions, and much of the remaining land was denuded and soaked with radiation, ghoulish legacies of the US attempt to learn more about the nuclear genie that was let out of the bottle to bomb Japan into submission in World War II.

Enewetak Atoll is well known for another form of force and fury, natural but more powerful than the nuclear tests. It lies in a region of the vast North Pacific Ocean, east of the Philippine Islands and southeast of Japan, that gives birth to more cyclones than any other region in the world. Twenty to thirty cyclones each year, out of a total of roughly one hundred born in the six cyclone-forming regions of the world, originate in the North Pacific Ocean.

In the waning days of September and arrival of October in 1962, the young weather disturbance was restless and on the move, ready to make a mark, ready to finally earn a name. It jogged to the northeast and left infancy behind; its gathering winds rushed to another stage, the clouds grew in size and number, and the winds reached 23 miles an hour, a sign of adolescence, a coherent, well-organized storm. By the morning of October 3, the rapidly intensifying weather force featured winds topping 38 miles an hour, which, by definition, made it a tropical storm and earned it the name Freda. The mass of clouds rotating around a low-pressure center stretched into the lower realms of the stratosphere some fifty thousand feet above the ocean surface. Traveling at 15 to 20 miles per hour, she breathed in, sucking warm moist tropical air from the ocean surface. She breathed out, exhaling cooler air aloft. By that afternoon, the winds reached a hurricane force of 80 miles per hour—74 miles per hour is the threshold for a typhoon or hurricane. Tropical Storm Freda was now Typhoon Freda.

The name "typhoon" applies to these weather monsters when they form in the northwest Pacific Ocean west of the international date line. The word is a botched pronunciation of the Chinese *ta-feng*, or "violent winds." In the Atlantic Ocean, Caribbean Sea, and eastern Pacific Ocean, these fierce, sprawling storms are known as hurricanes, the name a variation of *huracan*, which some Caribbean islanders use to describe the "god of all evil." Across the equator in the south Pacific, these same storms are sometimes called "willy-willies." Travel west into the Indian Ocean, and "cyclone" is the common name, and also the generic name for all typhoons and hurricanes. Cyclones are the biggest, most violent, and deadliest of all atmospheric disturbances spawned in the warm ocean waters of the tropics. Their fierce winds, torrential rains, and powerful sea surges make them the equivalent of meteorological bombs, which have exploded and killed some 1.9 million people over the past two centuries.

The practice of naming tropical cyclones dates back to the late-nineteenth-century Australian forecaster Clement Wragge, who named destructive typhoons after women he knew, or perhaps wanted to know. He was also known to name typhoons after politicians he held in low regard. Naming tropical cyclones became more common among military forecasters in World War II, and in 1953, the national weather bureau adopted the navy's practice of naming hurricanes born in the Atlantic Ocean after women. In 1970, the naming system was formalized, using women's names found in baby-naming books. However the women's rights movement was gaining momentum and, by 1978, the World Meteorological Organization, under intense lobbying pressure, agreed to include men's names, too. Each year, the first tropical cyclone of the year is given a name that starts with an "A." In 1962, Freda was the sixth typhoon of the season.

The tendency is to think of a tropical cyclone as the devil incarnate, but they serve a useful purpose, too. They deliver a semblance of order to the Earth's complicated weather system. Think of tropical cyclones as the great equalizers. They pull hot air away from the tropics. Without them, the areas around the equator would be an unbearable 20 degrees Fahrenheit warmer. They move that hot air toward the poles, balancing temperatures and precipitation around the globe, making more regions of the world habitable for people, plants, and animals. Beyond the death, destruction, and global weather control they provide, they've even changed the course of history. In 1281 AD, a typhoon that made landfall in Japan killed some hundred thousand Mongols, helping turn back an invasion. The Japanese called the typhoon *kamikaze*, and thanked their storm gods for intervening on their behalf.

Typhoons have demolished island villages, mainland communities, naval fleets, and explorer expeditions for centuries. But it wasn't until the middle of the twentieth century that the United States began stitching together a typhoon tracking system to protect their interests in the North Pacific Ocean. It took one of the most destructive typhoons in US naval history to get the typhoon monitoring program rolling.

On December 17 and 18, 1944, the US Navy Third Fleet, under the command of Admiral William "Bull" Halsey, took a break from air raids against Japanese airfields in the Philippines to refuel some three hundred miles east of Luzon. The admiral unwittingly sailed his ninety ships into the teeth of Typhoon Cobra, which was armed with winds approaching 150 miles per hour. Listing ships riding light and low on fuel sent aircraft plunging overboard. Three destroyers, 146 planes, and 790 men were lost, and eighteen other ships suffered major damage. It was a more crippling blow than Japanese forces might deliver. A navy court of inquiry ruled that Halsey had shown poor judgment, but he escaped sanctions. Halsey then sailed his fleet into another typhoon in late May of 1945, an action that claimed six lives and seventy-five planes. Another court of inquiry wanted Halsey relieved of his duties, but Admiral Chester Nimitz overruled the court.

Freda was five hundred miles west of Wake Island when navy weather observer Bruder and nine other members of his typhoon tracker squadron stationed on Guam climbed aboard their Lockheed Super Constellation EC-121, a triple-tailed aircraft equipped with four 3,400-horsepower engines and radar domes above and below the long, cigar-shaped fuselage that housed twelve thousand pounds of electronics. These airborne beasts of burden were designed to provide an airborne early warning system to track Soviet submarine movement and guard against missile attacks, as well as monitor the strength and movement of typhoons to determine

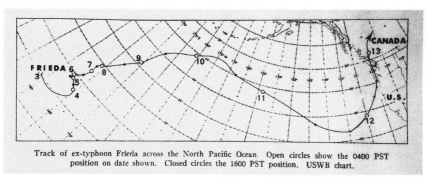

Track of ex-typhoon Frieda across the North Pacific Ocean. Open circles show the 0400 PST position on date shown. Closed circles the 1600 PST position. USWB chart.

Typhoon Freda (misspelled in graphic above) from birth to death. The line and numbers that appear across the North Pacific Ocean represent the storm track from October 3 to October 13, 1962. Map courtesy of US Weather Bureau.

whether they posed a threat to US and ally populations, military installations, and navy ships in the western Pacific.

Once in the air, Bruder went to work at his plotting table, relying on the plane's temperature probe, radar, and other instruments to measure wind speed, wind direction, barometric pressure, and temperature, sending data by teletype back to the Fleet Weather Central at Pearl Harbor, Hawai'i, every thirty minutes. This mission was straightforward: keep tabs on the storm and attempt to determine whether it would make landfall, or threaten US naval exercises at sea.

Typhoons in the North Pacific Ocean typically take one of three tracks. They may head west toward the Philippine Islands. The Asian archipelago is a common target, as witnessed November 7, 2013, when Typhoon Haiyan made landfall with winds of 190 miles per hour, a storm surge, and heavy winds, leaving a death toll of four thousand and damages to crops and property estimated between $6.5 billion and $15 billion. Other typhoons travel west, then curve north, placing Japan, Korea, and southeast China in the crosshairs. Then there are those with a more northerly track that either stay at sea or affect only small islands. Typhoon Freda took this third route.

Typhoon Freda, the eighth-strongest North Pacific typhoon of twenty-three spawned in 1962, continued to grow, looping to the northeast over the open waters of the largest ocean on earth at 25 degrees latitude and approaching 160 degrees longitude. By October 4, Typhoon Freda reached her peak wind velocity of 115 miles per hour, a category 3 typhoon, and her maximum low pressure of 948 millibars, another indicator of a powerful storm.

What a powerful, impressive sight it must have been, a convection engine with the moist warm ocean air rising vigorously within an eyewall of cumulonimbus clouds—those huge, unstable and vertical kings of the sky, boiling and towering over the top of the troposphere some fifty thousand feet high, dispensing showers, hail, thunder, and lightning. It's here in the eyewall that the winds and rain generated by a typhoon are strongest. Beyond the eyewall, several bands of cumuliform clouds form, extending several hundred miles in diameter. Bruder's dark gray plane kept a respectful distance, refueling on Wake and Midway Islands after the storm had passed each island without doing harm. When in the air the Lockheed cruised at 180 miles per hour at no more than ten thousand feet—the plane's cabin wasn't pressurized.

When typhoons appeared headed for land, the typhoon tracking crews would penetrate the storms to get a more precise reading of barometric pressure to determine the storm's potential destructiveness. That wasn't the case with Typhoon Freda, which stayed in the open ocean. "The higher intensity storms offered a bumpy ride until we went through the wall," Bruder said. "Inside the eye of the storm you could look down and the water surface would be perfectly calm. Exiting the storm we would go with the counter-rotating wind circulation and exit like being thrown from a slingshot."

An airplane penetrating a typhoon sounds like risky business. But Bruder said much of the potential turbulence is reduced by flying in along the same horizontal plane as the wind. He does, however, remember one particularly violent penetration during which everybody was buckled into their seats, except him. "I was the only one up reading instruments and plotting the information on the plotting table. I had my left hand on the table bar to secure me while I entered information with my right hand. All of a sudden the bottom dropped out and we descended maybe two thousand feet. The force flipped me in the air and my feet hit the parachutes in the ceiling. When we bottomed out I slammed the floor with my feet. I made a decision that was enough observing for that flight."

The storm slowly began to weaken on October 6, passing about four hundred miles northeast of Midway Island on October 7. Freda's march across the cooler waters of the northeast Pacific Ocean south of the Aleutian Islands would be the storm's undoing. Freda was a storm in transition, no longer fueled by the warm tropical ocean. Now the storm's still powerful, but diminishing, presence was a product of temperature differences between the tropics and the northern latitudes.

While Bruder and the other navy airmen were chasing Typhoon Freda across the ocean, Keith Clark was caught in the middle of the storm aboard the USS *Providence*, a guided missile cruiser and flagship of the Seventh Fleet. Clark was another teen who came of age during the peak of the Cold War and joined the US Navy Reserve as a high school junior in Olympia, Washington. He entered active duty upon high school graduation in 1961, in part to avoid the army draft. "Damned if I was going to dig foxholes," said Clark, as we talked over lunch at the Shipwreck Café along the shores of a muddy Puget Sound estuary near Olympia, Washington, fifty years and thousands of miles removed from his Southeast Asia tour of duty. Back then he was a twenty-year-old radar technician trained to help plot all ships and planes his 610-foot-long cruiser encountered at sea.

Since the USS *Providence* often had the admiral of the Seventh Fleet aboard, more time was spent sailing from port to port for goodwill tours and military strategy talks than practicing military maneuvers at sea. When typhoons struck, the ship was more likely to be at its home base— Yokosuka, Japan—or one of eleven other Japanese ports, or Singapore, or Manila in the Philippines, or Hong Kong, or carrying US ambassador Henry Cabot Lodge forty-six miles up the Saigon River to Saigon, Vietnam, for talks with the South Vietnamese leaders embroiled in guerrilla warfare with the Vietcong.

But on occasion, the *Providence* had run-ins with typhoons. Such was the case in early October 1962 during back-to-back collisions with Typhoon Emma and Typhoon Freda southeast of the Philippine Islands. The *Providence* spent three days battling the storms, said Clark, a mild-mannered, bespectacled man who looks the part of the retired budget analyst and Boy Scout leader that he is. The ship sailed through winds that couldn't be measured because the ship's wind gauges were pegged out at 150 knots. Clark's account is consistent with the October 5 peak wind velocity of Emma, which morphed into a category 5 super-typhoon.

It was the storm-whipped seas that Clark remembers most. The ship's deck was awash in waves and no one on the crew ventured outside for fear of being washed or blown overboard. The ship rolled several times, almost to the point of no return. "At the time we were all too young and too dumb to be worried," he said. "Our faith in the ship and its ability to handle adversity was unshaken." Clark's most vivid memory was a three-hour predawn shift in the watch station on the bridge some seventy-five feet above sea level. "I never saw the main deck the whole time because it was under water. A few waves broke over the bridge. I'd see the wall of water coming and just hang on." The ship's roller-coaster ride through the waves' peaks and valleys was heightened by an earlier blunder. The ship's engineering officer had failed to top off the fuel tanks before they set sail from the Philippine Islands, which added to the ship's instability. Back in port, the engineer was relieved of his duties, Clark recalled.

As Clark shared his memories of his time in the teeth of typhoons Emma and Freda, I couldn't help but think of all the ships lost at sea to hurricanes and typhoons over the centuries. The one thousand men aboard the USS *Providence* spent most of those three days lashed to their bunks or working in the inner sanctum of the ship. But imagine life aboard the early sailing ships, where the men were exposed to the full fury of the storms.

In his classic tale *Typhoon*, Joseph Conrad captures the horror experienced by the crew of the steamer *Nan-shan* as it runs head-on into a nineteenth-century typhoon in the China Sea.

It was something formidable and swift, like the sudden smashing of a vial of wrath. It seemed to explode all around the ship with an overpowering concussion and rush of great waters, as if an immense dam had been blown up to windward. In an instant the men lost touch of each other. This is the disintegrating power of a great wind: it isolates one from one's kind. An earthquake, a landslip, an avalanche, overtake a man incidentally as it were—without passion. A furious gale attacks him like a personal enemy, tries to grasp his limbs, fastens upon his mind, seeks to rout his very spirit out of him.

No doubt, the double-whammy typhoons grabbed the attention of USS *Providence* crewmembers. But bad weather at sea wasn't much talked about, or dwelled upon, Clark noted. A case in point: the three-hundred-page yearbook describing the *Providence*'s flagship cruise in the North Pacific from 1962 to 1964 offers not a word about stormy weather at sea. It does devote fourteen pages to a "Crossing the Line" ritual—make that a physical hazing—of sailors crossing the equator for the first time.

Then there's the sheer size of the sea in a cyclone. Oceanographers theorize that hurricane-force winds could pile up seas into waves topping out at 220 feet, the equivalent of a twenty-two-story building. Nothing like it has ever been witnessed, but given the vast ocean spaces where cyclones roam, it's possible these monster waves of unimaginable size have occurred without notice. The size of waves that have been recorded is frightening enough. Hurricane Ivan, a category 3 tempest with wind speeds topping 115 miles per hour in the Gulf of Mexico in 2004, generated waves estimated at 130 feet tall. In September of 1995, the *Queen Elizabeth 2*, a luxury liner sailing from the French seaport of Cherbourg to New York City, tried to avoid the wrath of Hurricane Luis, but still encountered a ninety-five-foot-high wave and repeating waves up to sixty feet tall.

On October 9, Typhoon Freda was leaving the warmer waters of the western Pacific, but still packed a punch. This is no typical typhoon, Bruder thought to himself. As the day turned to early evening, the typhoon had traveled far enough across the ocean for Bruder to see it framed in the radar with Alaska's Aleutian Islands chain to the north and the West Coast of North America to the south. This was the farthest any of Bruder's flights took him across the Pacific Ocean from his home base in Guam. Bruder sent a message back to Fleet Weather Central at Pearl Harbor about 6 p.m., telling the duty officer there that Freda was still a powerful storm. One option would be for the crew to land at Adak, Alaska, refuel, rest, and continue to track the storm the next day. Clothed in only lightweight tropical flight suits, the crew did not look forward to landing in freezing, snowy Adak. As it turned out, their worries about having to confront Alaskan weather were for naught.

"I told them this storm is going to hit the West Coast of the United States and cause major damage," Bruder recalled. How did his superiors respond? "Go back to Guam. The storm is going to cool down and dissipate," the message came back. The twenty-year-old navy aerographer did what he was told. He didn't question authority. He followed orders. The plane turned around and began the two-day flight back to Guam. On the return trip the plane was struck by lightning. It rocked the whole plane, but did no internal damage, Bruder recalled. Back on Guam the next morning, the crew found a six-inch hole in the plane's vertical stabilizer.

As Bruder's crew flew back to Guam, Typhoon Freda morphed into an extratropical, midlatitude cyclone, which means it switched energy sources from the latent heat in the warm tropical waters to the air pressure and temperature differentials found between the tropics and the northern latitudes. On October 10, it continued its amazing journey across the Pacific Ocean, now entrained in the jet stream. By the evening of October 10, Freda was downgraded to a tropical storm, crossing the international date line at 180 degrees longitude and sticking close to 45 degrees latitude. Still a well-defined storm, the former Typhoon Freda's sustained winds had dropped to 45 miles per hour.

On October 13, the day before his twenty-first birthday, Bruder was back in the concrete barracks on the Guam navy base. He occupied a lower bunk for one particular reason: At night, geckos—small, nocturnal lizards—would crawl on the ceiling and occasionally drop onto upper-bunk occupants. Bruder picked up the *Pacific Stars and Stripes* newspaper. The big, bold headline on the front page screamed out the news of an October 12 Columbus Day windstorm that had devastated the West Coast. Bruder knew that Typhoon Freda was at least partially to blame for the swath of death and destruction that stretched from Northern California to British Columbia. "Well, I said to myself—we told them it was still a full-blown storm," he said. "There was nothing else I could do."

Bruder and his crewmates returned to their more mundane land-based routines. Reveille at 7 a.m., breakfast in the chow hall before climbing aboard a forty-foot-long trailer called the "cattle car" to cross the runway to the duty station for training. Later in the day there was runway guard duty, planes to clean, and equipment to test and maintain. A week after tracking Typhoon Freda, Bruder was on the wing of his plane, enjoying the sunshine and scrubbing away, when the crew was approached by navy office staff who told them the tours of duty of all personnel were extended indefinitely because of the "Cuban crisis." It was Bruder's first hint of what would become the Cuban Missile Crisis.

3
Countdown to Calamity

As the unsettled remains of Typhoon Freda moved southeasterly in the jet stream south of Alaska on October 10, a strong storm was forming eight hundred miles west of San Francisco, shaped by a mass of cold air spawned in the northern Gulf of Alaska that moved south, mating with warm, moist air from the southern latitudes. The conditions were ripe for the type of wind and rainstorm that makes landfall in the Pacific Northwest two or three times every decade. Think of the ferocious storm that moved quickly into Northern California and Southern Oregon in the early morning hours, Thursday, October 11, as the opening act to the Columbus Day Storm main event October 12.

The weather report issued Wednesday night for Thursday by the US Weather Bureau in San Francisco called for a 90 percent chance of rain and winds 25 to 35 miles per hour. But the gale force winds doubled in velocity from their predicted strength, and the rain that began on Thursday was torrential, just the beginning of an onslaught that topped ten inches in many places in Northern California before the weekend was over, marking it the worst Bay Area downpour since the early 1900s. The deluge triggering fatal car crashes, deadly mudslides, and widespread flooding. The Thursday storm registered severe winds up and down the West Coast from the Bay Area to Washington State, including 80 miles per hour at Mount Tamalpais, which rises above the Marin Hills north of San Francisco. In Astoria, Oregon, at the mouth of the Columbia River, the winds topped out at 62 miles per hour.

On Thursday, some fifty fishing boats from the Eureka, California, commercial fleet in Northern California had left behind the relative safety of the Humboldt Bay estuary, that grand confluence of sea meeting river at the gateway to redwood country, home to the tallest trees in the world. The fish-seeking vessels knifed their way westward into an unwelcoming ocean, defined that day by 60-mile-per-hour winds and thirty-foot seas. The sea's bounty of Dover sole, salmon, shrimp, and tuna, which was the fish of interest that day, was the last thing on the minds of skippers and their crews as they tried to keep their boats afloat.

The furious winds and seas Thursday morning triggered May Day calls from several fishing ships and responses from three Coast Guard cutters from Eureka, Fort Bragg, and San Francisco and two smaller Coast Guard lifeboats stationed in Humboldt Bay. The eighty-three-foot Coast Guard cutter *Avoyel* from Fort Bragg reached the sinking *Signe F* in the nick of time, placing it under tow several miles southwest of Eureka. The boat *Jean Ellen* sank some fifteen miles west of Eureka, but not before the two men aboard, skipper Douglas Fearon and crewmember Al Vierra, were rescued by skipper Richard Haman and his crewmember Bruce Campbell aboard the Eureka-based *Elsinore*. Four years earlier in local waters, Haman had pulled a man, woman, and dog off the sinking boat *Colleen*. Details of the dramatic Thursday rescue were sketchy. But a photo of the four men, taken around 11 a.m. at the Wojcek & Swiss Fish Company dock, showed Fearon with his arms slung across the shoulders of Haman and Vierra, his face fixed in a state of shock and loss. Haman, a handsome, tall man of the sea, shows a deep, reflective look in his eyes while two crewmembers look on with broad grins of relief.

An initial report of a crewmember washed off the decks of the fishing boat *Mary Elaine* proved erroneous. But as the fishing vessels limped back to harbor, the thirty-eight-foot Eureka troller, *Star*, was missing, and its sole occupant, John Essex, forty, was feared lost at sea. To the north, the Coast Guard struggled for hours to bring a floundering forty-two-foot fishing boat called *Petrel* to safe harbor in Depoe Bay on the Central Oregon coast. The boat's captain, Paul DeBellow of Eugene, Oregon, had left the Columbia River early Thursday morning, bound for Newport, Oregon. At 7:55 a.m., he radioed the Coast Guard that he was in trouble, bucking high seas and southerly winds. First, a thirty-six-foot Coast Guard rescue boat tried to tow the fishing vessel to safe harbor on the north-central Oregon coast, but the winds were too strong. A second, fifty-two-foot Coast Guard ship from Newport had to be called in to action to help complete the rescue.

The Thursday storm directed much of its fury at the coastal village of Gold Beach in Southern Oregon, about fifty miles north of the California border. It's an isolated coastal stretch where the Rogue River, federally protected today under the Wild and Scenic Rivers Act, flows from the Siskiyou Mountains and pours into the Pacific Ocean. This stretch of coastline is known as the banana belt of Oregon. Temperatures topping 70 degrees Fahrenheit occur sometimes in the winter on the Southern Oregon coast, which also holds the record highs for Oregon every month November through March. This weather anomaly can be traced to occasional offshore flows of warmer compressed air off the slopes of the mountains.

In October, however, winds flow mostly from the south and west, and October 11 was no exception. Gold Beach was at the epicenter for Thursday storm damage. In the early morning hours, the winds tore the roof off the Riley Creek Elementary School and knocked down the walls of the gymnasium, school kitchen, multipurpose room, and boiler room. "Thank God it wasn't two hours later," a school official said that day. At 9 a.m., a school assembly for four hundred students was scheduled in the gymnasium. The Thursday windstorm also damaged thirteen homes in Gold Beach and took dead aim at Pacific Trailer Mart, hurtling through the air nine trailers—some of them fifty-five feet long—and stacking others up like cordwood. Property damage in Gold Beach was pegged at more than $7 million.

Five deaths in Northern California, Oregon, and Washington were blamed on the run-up to the big storm Friday. One of the victims was thirty-eight-year-old Roger Whitman, who lived in Bothell, Washington, a suburb of Seattle. A meter reader for Puget Sound Power and Light Company, Whitman's work truck was smashed by a falling tree Thursday near North Bend, Washington, in the foothills of the Cascades. "He was the only fatality in King County," recalled his son, Bill Whitman, who was a first-grade student at the time. "Among my memories are those of my weeping mother, concerned neighbors, and hearing my father's name mentioned on the *Huntley-Brinkley Report* on our old black-and-white TV."

Whitman said his mom and dad had just leased space to open a new business in Kenmore, Washington, another Seattle suburb. His father's death forced his mother to sell their home and move the family to a modest mobile home on the banks of the Pilchuck River in rural Snohomish County north of Seattle. "Although the storm significantly altered her life, she rarely spoke of it," Whitman said. "Late in her life, after suffering a stroke, she shared with me that she felt her duty was to put the past behind and do the best she could with what she had." Whitman said he treasured that thought and his mother's ability to endure and adapt after the storm tragedy.

Early Thursday morning, the remnants of Typhoon Freda were on the move about nine hundred miles south of Dutch Harbor, Alaska, slipping around a high-pressure system that covered the mid-Pacific. Its barometric pressure reading was 29.59, fairly typical for the season in that part of the Pacific. By nightfall Thursday the torrential rains and winds had subsided. The Friday weather forecast, issued 9 p.m., Thursday, called for rain, heavy at times, and gale force winds reaching 30 to 45 miles per hour. It was a stormy forecast, but nothing compared to what was about to happen.

Late Thursday night a low-pressure system that was the remains of Typhoon Freda reached the offshore stomping grounds of the Thursday storm, with a storm center about sixty miles west-southwest of Fort Bragg in Northern California. Most weather observers assumed that the latent energy from the first storm was probably already spent. The weather reports received by the US Weather Bureau from ships at sea 11 p.m., October 11, were nothing to cause alarm. The barometric pressure at the storm's center was 29.21, well within the range of normal fall weather and just .09 inches lower than six hours earlier. The next set of coded weather reports from ships at sea was not due for another six hours.

Ships at sea were the eyes and ears for coastal weather forecasters in 1962. Leading the way were sixteen World War II cargo ships that were retro-fitted, from 1954 to 1959, with radar and stationed, five at a time, three hundred to four hundred miles off the West and East Coasts to watch for Russian long-range bombers. They had a secondary responsibility, relaying weather information to the weather bureau to help the federal agency shape its weather forecasts. These reports from ships at sea were especially crucial when big storms were brewing off the Pacific and Atlantic coasts, storms like the Columbus Day severe weather of October 1962.

At 5 a.m., Friday, October 12, a navy radar picket ship off the Northern California coast reported a rapid drop in barometric pressure and winds up to 59 miles per hour. The weather bureau took the information to heart and issued gale warnings of winds up to 54 miles per hour Friday for coastal areas from Northern California through Washington. It was starting to look like a repeat of Thursday's storm. Aviation forecaster Jim Holcomb was on duty Friday morning at the weather bureau in Portland, reporting for work at 7:00. Holcomb, twenty-eight, and a native of Seattle, Washington, cultivated an interest in weather as a young boy peering at the night sky through a telescope and trudging through knee-deep snow during the Seattle blizzard of January 1950. After graduating from the University of Washington with a degree in atmospheric sciences, Holcomb went to work for the US Weather Bureau at the Los Angeles International Airport. He grew bored quickly. "I was interested in exciting weather, and Los Angeles weather was like the movie *Groundhog Day*," Holcomb said of the uninterrupted stretch of sunny, blue-sky days. He jumped at a chance to transfer to the Portland weather bureau in 1961.

Soon after his October 12 shift started, Holcomb entered the weather bureau room that housed the noisy, paper-consuming teletype machines— one for receiving ship reports, one for weather observations from other land stations, and a third for transmitting Portland forecasts to forecast users and other weather bureaus. The teletype machine dedicated to ship reports was spitting out the 8 a.m. transmittal from a picket ship 340 miles west of Fort Bragg, California. It proved to be the single most important, and foreboding, report weather forecasters received that morning. The ship was stationed where the center of the newly formed storm had been at 4 a.m. The winds were pummeling the ship at 92 miles per hour, and the barometric pressure had fallen 0.66 inches in a three-hour period, an astonishing pressure drop and a clear sign the storm was growing into a meteorological bomb capable of unleashing hurricane-force winds. "When I saw the teletype, I could hardly believe my eyes" he said. "I'd never seen a barometric pressure drop of that magnitude before, and I've never seen one like it after."

Holcomb took the information to lead forecaster Dub Yates, who was putting the finishing touches on the morning weather forecast. Could it be an error? Yates asked. There was no way to verify the numbers, but the wind and pressure readings were too consequential to just ignore, Holcomb said. With the morning weather observation from the picket ship in hand, Yates reworked the weather map and concluded the center of the intensifying storm at 9 a.m. was about 205 miles west of Eureka, California, not far from where the coastal town's fishing fleet had been battered by winds the day before. Forecasters were puzzled by the lack of wind along the Northern California coast—only 5 miles per hour at Crescent City. However, the rapid drop in barometric pressure reported by the radar picket ship convinced them to issue this weather advisory at 10:10 a.m.:

> A very deep vigorous storm that is the remnant of a Pacific Ocean typhoon storm is moving in a northeasterly direction and expected to be centered over the chan [channel] near the mouth of the Columbia River this afternoon. Ship reports near the storm center indicate that it is a very dangerous storm. Whole gale warnings [just short of a hurricane wind warning] are displayed on the north California coast and gale warnings on the Oregon and Washington coast are expected to be changed to whole gale warnings in the next few hours. Precipitation amounts are expected to be very heavy over Western Oregon and heavy over Western Washington during the next 24 hours. This advisory weather warning is recommended for immediate and frequent broadcast in Western Oregon.

The wind warning that followed at 10:40 a.m. was a point of contention between Holcomb and Yates. Holcomb wanted to issue a wind warning for Portland-area airports, suggesting winds of 40 knots later Friday afternoon, gusting to 60 knots. "Dub said that was a little strong, a little scary," Holcomb recalled. The aviation and general wind warning was toned down a little bit. The wind warning issued from the Portland station at 10:40 a.m. read, "East to southeast winds increasing to 20 to 30 knots early this afternoon shifting to south to southwest 20 to 40 knots with gusts to 60 knots late afternoon. Winds gradually diminishing this evening."

The prediction of winds reaching 60 knots in the Portland area was the highest velocity wind warning ever issued by the weather bureau at the Portland airport. The forecast, equal to a wind gust of 69 miles per hour, was still shy of the 72-mile-per-hour record gust logged at the airport years ago, Holcomb said. There was a reason for that, Holcomb explained: the weather bureau didn't want its forecasters trying to predict record events.

It was not an easy forecast to make, recalled Jack Capell, the Portland, Oregon, television weather reporter, in his storm postmortem called *The Terrible Tempest of the Twelfth*. For instance, at the weather stations closest to the storm at 10 a.m.—Crescent City, California, and Brookings, Oregon—winds from the south-southeast hovered around 18 to 20 miles per hour, brisk but nothing out of the ordinary along the coast. "It was no wonder that the forecasters who issued the warnings for a windstorm of almost unprecedented violence felt some apprehension that they had possibly needlessly alarmed the public," Capell wrote in his recount of the storm. "There was still the chance that the navy picket ship had suffered some gross error through its radio and teletype chain of transmissions."

But crewmembers from navy picket ships ensnared in the storm amid forty- to fifty-foot swells and winds topping 90 miles per hour knew their reports rang true. "I spent three years at sea and that was the worst storm we ever had," said John Hubbard, a retired boilermaker from Midland, Michigan, who was a deckhand on watch that morning on the bridge of the USS *Tracer*. Hubbard said all on board wondered how much more pounding the flat-bottomed ship could take before it broke apart. "We'd crest a wave and the ship's screws [propeller] would come out of the water," he said. "I'd never seen swells that big. I was scared shitless. I still thank God and the captain of the ship that we made it."

At one point, a giant wave swept over the port side of the bridge wing, dousing the port and starboard lookouts in a crashing wall of frothy seawater. Hubbard feared they had been swept overboard. "We hurried and opened the port side watertight door and the lookout was completely wet and wide-eyed," Hubbard said. "On the starboard side the lookout was hulked down in the corner next to the door, white as a ghost."

The radar picket ships were part of the North American Air Defense (NORAD), which was formed by US and Canadian armed forces in 1958 to try to protect the North American continent from enemy air attacks. The US Air Force maintained overall command of the operation and contributed land-based radar stations, fixed offshore sites, and early warning planes. The navy supplied the picket ships and other aircraft, and the army installed Nike sites around the country with antiaircraft missiles, just in case a Soviet bomber, or bombers, slipped through the other lines of defense.

Time at sea aboard these old Liberty ships, which were more than four hundred feet long, with crews of 150 men, could be monotonous and span 250 days a year. But the navy went out of its way to boost morale with a variety of activities and amenities, including onboard basketball courts, movie theaters, archery ranges, spacious sleeping quarters—at least for a ship—and food befitting a civilian restaurant. Tuna fishing derbies, sanctioned by the navy and the US Fish and Wildlife Service, pitted ships' crews against each other during extended stays at sea. After a month-long tour of duty in the fall of 1964, the USS *Interceptor* returned to home port at Treasure Island in San Francisco. The crew hadn't caught any enemy aircraft in its radar net, but it had pulled in 421 albacore tuna weighing a combined 3,852 pounds, making them the fishing champions of the Pacific Fleet.

Fishing and other leisurely pursuits was the last thing on the minds of the radar picket ship crews October 11 and 12 as they battled to keep their ships afloat. The USS *Picket* was west of Portland, Oregon, headed back to Treasure Island when the storms struck. The ship had been relieved of duty off the west coast of Vancouver Island, British Columbia, by the USS *Watchman* on October 9. The trip home was a harrowing one, assistant navigator Carl Coad recalled: "Fifty-five foot seas were washing over the bridge and pilot house, and the ship was rolling 45 degrees both ways," Coad said. "We were defenseless." Coad, a retired construction company estimator living in Olathe, Kansas, in the winter of 2014, said the ship, which had a top speed of 10.5 knots per hour, was headed directly into the wind and getting pushed backward about 1.5 knots per hour. The roller-coaster ride up and down the ocean waves cracked the ship's hull. "If we'd been in that storm much longer, we would have broken in two," he said.

Any lingering doubt about the veracity of the radar picket ship weather reports were swept away just before 11 a.m., when a Brazilian cargo ship sent a brief, coded message describing 63-mile-per-hour winds from the south-southeast, about sixty miles southwest of Cape Blanco, an exposed headland on the southwest Oregon coast that juts out 1.5 miles into the ocean, making it the westernmost location in the state. An hour later, the Brazilian ship's crew felt compelled to send a valuable update to the weather bureau: winds had increased to 83 miles per hour out of the southeast in a steady rain.

Cape Blanco, home to a United States Coast Guard LORAN (long-range navigation) Station from 1945 to 1980, lived up to his reputation as the windiest place on the Oregon coast Friday. The station's wind-damaged anemometer registered 145 miles per hour, and there were unofficial estimates of 179-mile-per-hour wind gusts at the peak of the storm. The weather bureau had no knowledge of what was happening at Cape Blanco.

Wind observations there were missing from the 10 a.m. roundup, probably because of power outages at the Coast Guard station.

Oregon native Chris Percival was among the dozen or so "Coasties" assigned to the Cape Blanco station when the storm struck. The freshly minted petty officer third class and electronics technician described coming off graveyard duty about 8 a.m. from the signal tower building, where the LORAN equipment was stationed, and trying to return to the bachelors' quarters. "I was immediately knocked down by the wind and had to crawl on my hands and knees back to the barracks," he said. "At one point the wind picked me up and wrapped me around a telephone pole."

Bruised and battered, but spared any broken bones, Percival remembered huddling on the floor of the barracks with the wind whistling through the broken-out windows. Gravel and rocks flew through the air, breaking the beacon light lens on the lighthouse some eighty-five feet above the ground. The station's radio beacon tower, consisting of one-inch-square steel latticework, twelve feet square and 125 feet tall, anchored by two-foot-square concrete blocks, collapsed in the wind.

By the noon hour, the storm center was believed to be about 140 miles west of Crescent City, and noon wind gusts there had reached 46 miles per hour, with a still dangerously low barometric pressure reading of 28.67 inches. Over the next two hours, wind velocities continued to climb—50, 60, 70 miles per hour in Northern California and Southern Oregon. At Requa Air Force Station, perched on a bluff overlooking the Pacific Ocean at the mouth of the Klamath River in Northern California's Del Norte County, a wind gauge registered 112 miles per hour Friday morning as the winds tore roofs off buildings at the military base. Twenty miles north, in the coastal town of Crescent City, half the homes and businesses in the city suffered damage in the form of broken windows and blown-away television antennas, roofs, and fences, according to Crescent City police. Most major roads in the county were blocked by fallen trees, including huge coastal redwood trees that succumbed to the wind.

In Shasta County, more than a hundred miles inland from the Northern California coast, wind racing across Shasta Lake created waves six feet high. The rain-soaked tail of the storm lingered over Northern California, triggering river flooding, lethal landslides, bridge and railroad line washouts, and fatal car accidents. As the main body of the storm spun north into southern Oregon, Friday's rainfall total in the isolated inland communities of Northern California topped three inches in many places, including six inches in Garberville, a city of a thousand people in Humboldt County, and a staggering twelve inches at Squaw Creek at the north end of Shasta Lake.

The storm crossed into Southern Oregon in the early afternoon, striking coastal communities that had had no time to recover from Thursday's destructive blow. In Brookings, Oregon, a coastal community just across the California border, winds clawed at buildings, trees, and power lines, collapsing a school bus barn and sending a huge roof beam from the Crest Motel flying into a nearby duplex, slicing it in half. In Port Orford, some sixty miles north of Brookings on coastal Highway 101, winds tore apart sections of Pacific High School about one hour after students had been released from school. Two dozen homes were destroyed and 111 others were damaged.

Weather bureau meteorologists in the Portland office issued this updated weather advisory at 1 p.m., Friday, about the time the storm damage in Southern Oregon began to mount:

> The storm that intensified off the California coast early this morning has been moving in a northeasterly direction at near 50 knots. It is now about 150 miles west of Coos Bay or North Bend, Oregon and will continue northward to about 100 miles west of the mouth of the Columbia River by mid to late afternoon. The storm should slow down some but is expected to be near Vancouver Island by early Saturday.
>
> This is an intense storm and whole gale warnings are displayed on the coast from northern California to the northwest tip of Washington and through the Strait of Juan de Fuca. Gale warnings will be ordered on the island waters [Puget Sound] of Western Washington within the next two hours. Precipitation is expected to be heavy over all of Western Washington and Western Oregon.

The advisory noted that Saturday was the opening of hunting season in Washington: "Hunters are urged to use caution in selecting camping sites in regard to falling snags and trees in exposed locations on the coast and in the mountains."

If the weather advisory seemed conservative, compared to what was happening on the ground, there was a reason, Capell said in his postmortem on the storm. "Even as late as 1 p.m., no Oregon station on the teletype circuit showed any particularly strong winds with the single exception of the 3,800-foot peak in southwest Oregon called Sexton Summit. They had southwest [wind] to 47 [miles per hour]. It wasn't until 2 p.m. that southwest Oregon began to feel the first substantial effects of the blow." The relentless storm marched north up the coast and into the Willamette Valley, turning Oregon's most populated area into ground zero for the storm.

Holcomb relived his most eventful day as a weather forecaster on October 12, 2016, fifty-four years removed from the Columbus Day Storm. He was back in Portland, speaking to a dozen senior citizens enrolled in a continuing education class offered by the Senior Studies Institute of Portland Community College. The title of his talk was "Forecasting the Columbus Day Storm." The lanky, soft-spoken meteorologist painted a picture of weather forecasting in 1962 that was more art than science. Holcomb and his colleagues had the challenging job of trying to identify and track weather systems that typically move from west to east in the midlatitudes. That means West Coast weather is a product of what is happening over the vast Pacific Ocean, the world's largest water body, covering more than 30 percent of the earth's surface.

"Without the ships at sea that day, we had next to nothing to work with," Holcomb recalled. Other tools at their disposal were limited. The National Aeronautics and Space Administration launched its first experimental weather satellite in 1960, part of the Television Infrared Observation Satellite (TIROS) program. The fifth TIROS satellite was in orbit when the Columbus Day Storm struck, but was providing only spotty, daytime-only photos of clouds covering whatever part of the earth's surface it was passing over daily. Holcomb said they typically received one snapshot of the eastern Pacific Ocean each day, and the photo October 12 was not memorable.

Compare that to present-day US satellite coverage, which includes continuous, day and night coverage of the Pacific and Atlantic Oceans and the United States from two geostationary satellites, which maintain fixed locations by orbiting above the equator at the same speed as the Earth's orbit. Those satellites are complemented by two lower-orbiting weather satellites that move north to south over the North and South Poles, providing images of everywhere on Earth every six hours. Radiosondes, small weather stations sent aloft in helium or hydrogen-filled balloons to radio back to land stations weather data including temperature, humidity, and barometric pressure, were commonplace in 1962, but they weren't dispatched over the ocean. Holcomb and his colleagues did not have the benefit of Doppler radar, which today's meteorologists use to detect the speed and direction of wind and precipitation inside storms.

With the development of digital computers in the 1950s, weather forecasters stepped into the world of numerical weather predictions, using mathematics and the laws of physics to predict the weather. In 1962, the computer-generated weather forecasts were simple and imprecise, lacking the millions of data observations that feed the mathematical equations used by today's high-speed computers. The endless stream of weather data from satellites, weather buoys, ships, airplanes, radiosondes, and radar, coupled with ever-increasing computer power, has propelled meteorologists in recent years into the world of ensemble forecasting. Instead of running a single forecast with one set of data, they run multiple forecasts with differing inputs of data to produce a set of forecasts that indicate the range of possible weather to come in the days ahead. Ensemble forecasting can be applied to more than one computer model to develop a weather forecast, something called multimodal ensemble forecasting.

Holcomb and other meteorologists working in the early 1960s had confidence in their ability to predict the weather with some degree of certainty about twenty-four hours in advance. But the ferocity of the Columbus Day Storm winds outstripped their forecasting expertise. "At that time, we didn't have anything to tell us what the winds would be," he said fifty-four years later. "We were kind of pulling numbers out of a hat." If a replica of the Columbus Day Storm formed today, forecasters could provide the public with up to five days of advance warning, Holcomb said.

"You can't give the weather forecasters at the time of the Columbus Day Storm a poor grade because of the lack of tools they had to work with," UW atmospheric sciences professor Cliff Mass said. "The big story is how much weather forecasting has improved. The triumph of weather prediction is one of the grand achievements of our species." Despite all the advances in weather forecasting, meteorologists remain limited in their ability to predict next week's weather. The atmosphere is a chaotic system full of sudden changes and surprises. One of the first to recognize this was Massachusetts Institute of Technology professor Edward Lorenz, the father of chaos theory and the butterfly effect. It was Lorenz, in the early 1960s, just as weather forecasting was moving from childhood to adolescence, who threw cold water on the future of long-range weather forecasting with his "chaos theory," which showed that minute differences in a dynamic system such as the atmosphere today can lead to large effects in the future. The theory helps explain how a seemingly insignificant change in a mathematical model of atmospheric conditions can send a long-term weather forecast off the rails. Lorenz popularized chaos theory with an academic paper he wrote in 1972 called "Predictability: Does the Flap of a Butterfly's Wings in Brazil Set Off a Tornado in Texas?" While the question sounds tongue-in-cheek, the implications are real: those flapping butterfly wings take the certainty out of the long-range weather forecast equation.

While most residents in the sparsely populated, heavily timbered far west corner of the country just went about their business that fateful Friday afternoon, others took action to protect themselves and their charges, if they had a barometer to tell them the air pressure was plunging like a rock. Al Spady, a big, raw-boned logger and bus driver for Waldport High School on the Oregon coast, was eating lunch at home when he noticed his barometer's needle plummet. He rushed to school, stormed into the gymnasium and tromped onto the stage in his logging boots and plaid shirt, interrupting a pep rally in advance of a Friday night football game. "All you kids who ride my bus—get on it," he said in the loud voice of a logger half-deaf from the roar of chain saws in the woods. "All hell's gonna break loose, and we're going home."

By the time school officials received Coast Guard official notice of a major windstorm moving up the Oregon coast south of Waldport, Spady was already on the road, driving south on US Highway 101, dropping some students off in Yachats eight miles south of Waldport before heading east on Yachats River Road. All his charges made it home safely. A second bus that got a late jump on the storm ended up stranded on Alsea Road east of Waldport, pinned down by fallen trees. "Al, although probably the least educated, was by far the smartest," recalled Yachats fire chief Frankie Petrick, a high school junior in 1962. "He didn't need to know all the details about why the barometer had dropped. He just knew he had to get the heck out of there."

There's no record of what Spady's barometer registered when he sounded the alarm. But earlier in the day, the crew on a navy radar ship off the coast of Northern California caught in the oncoming storm watched the ship's barometer plunge 0.66 inches in three hours, to an alarming 28.41 inches. By comparison, sea-level barometer readings typically range from 29.70 to 30.20 with an average reading of 29.92. Any barometric pressure reading below 29 inches at a storm's center suggests a major storm. The world's lowest recorded barometric pressure was 25.69 at the center of Typhoon Tip in the western Pacific Ocean in 1979. That, too, was on Columbus Day—October 12.

The descending barometer readings that began off the Northern California coast and continued along the south-central Oregon coast also caught the eye of forecasters at the US Weather Bureau station in Portland, Oregon, some hundred miles north of Spady's rural coastal home. Portland television weather forecaster Jack Capell received a call at his northeast Portland home around noon from US Weather Bureau meteorologist Glen Boire. "Jack, you might want to get out here and take a look at what we're getting on the teletype," Capell recalled years later. "He said the barometer was dropping like crazy." Capell said goodbye to his wife, Sylvia, and four-month-old son, John, and headed out the door about three hours earlier than his normal workday visits to the weather bureau, where he shaped his evening weather forecasts for his television and radio audiences. It would be more than thirteen hours before he would see his wife and child again.

Michael Korte was a twelve-year-old living in Portland, hanging out at his friend's house the afternoon of October 12, 1962. The friend's father, Jim Haner, was a longshoreman and former seaman who had barometers stationed all over the home. Korte described in lay person's terms what he saw. "The barometers started doing loop-de-loop in unison—all of them." For the elder Haner, the only other sudden barometric pressure drop he'd witnessed was years ago at sea in the South Pacific, caught in a typhoon. He called Korte's mother, who worked at a bank in Portland, and urged a woman he didn't even know to leave work and head for the safety of home. It was 3:30 p.m., about two hours before the ferocious windstorm struck Portland.

Bob Dernedde remembered physics professor George Kurtz at Pacific University in Forest Grove west of Portland bursting into his campus office Friday afternoon, clutching a barometric pressure chart and claiming the campus west of Portland was in for a really big blow. The chart was the product of an aneroid barometer, which was invented by French scientist Lucien Vidi in 1840 and replaced the use of mercury with a metal vacuum disc, mechanical arms, and a pointer that marked changes in barometric pressure on a rotating paper drum. "We were already experiencing heavy winds, so we went to the [college] president and urged him to close the campus," Dernedde said. "He agreed and we went from office to office and building to building, telling everyone to go home or back to their dorms. It worked."

Jennifer White was a fourteen-year-old girl, walking home from school in the north-central coastal town of Newport, Oregon, Friday afternoon, October 12. The wind was gusting but it was warm, not the usual southwest storm she was used to. Her parents, Burnelle and Teddy White, operated a weather station for the US Weather Bureau, and phoned their observations to the Salem weather office four to six times a day. "When I got home, I asked mom about the warm wind," White said. "She said she had just talked to Salem, and they said it was a small typhoon, nothing to worry about." Thirty minutes later, Teddy White was adding paper to the bottom of the barograph, trying to keep up with the deepest drop in barometric pressure she had ever seen—down to 28.50 inches. The winds were gusting to 80 miles per hour, a portent of what was about to strike Newport.

The barometer was an unparalleled beacon of stormy weather, a mighty, lifesaving tool that provided the best clue to everyday citizens and meteorologists alike that a windstorm like no other was about to strike the Pacific Northwest on October 12, 1962. Those who were attuned to the significance of a sudden drop in barometric pressure had seventeenth-century Italian inventor and mathematician Evangelista Torricelli to thank. "We live submerged at the bottom of a sea of elemental air," Torricelli wrote in a June 11, 1644, letter he sent from Florence, Italy, to his friend and fellow scientist Michelangelo Ricci in Rome. Torricelli, a Renaissance-era protégé of Galileo Galilei, was enthused about his discovery in the spring of 1644 that the Earth's atmosphere has weight—later calculated at a mind-numbing 5.5 quadrillion tons—and that wind was air in motion, flowing from areas of high pressure to areas of low pressure. He helped put to rest the two-thousand-year-old theory advanced by Aristotle, the Greek philosopher and scientist, who defined winds as exhalations of a dry, vaporous element emitted from holes and fissures in the Earth.

Torricelli's breakthrough discoveries included an experiment in Florence in which he placed mercury in a meter-long tube sealed at one end. He placed his finger over the open end of the tube then tipped the tube upside down into a basin. When he removed his finger, not all of the quicksilver flowed into the basin; some remained in the tube. Torricelli surmised that atmospheric pressure—the weight of air bearing down—was keeping the mercury in the tube. His theory was reinforced as he watched the mercury level rise and fall as the weather changed from fair to stormy.

Torricelli, also a brilliant mathematician and expert lens crafter for microscopes and telescopes, had just uncovered the underlying principles of the barometer, the tool that measures changes in air pressure, the tool that helps forecast changes in weather: fair weather as the barometer gauge rises and stormy weather as it falls. He shared his newfound knowledge only with Ricci. He feared too much talk would lead him down Galileo's path, which included Galileo's 1633 inquisition by the Roman Catholic Church, his house arrest, and the threats of torture and burning at the stake for proclaiming that the earth rotated around the sun. Torricelli used vacuums in his experiments, and the church believed, since God was everywhere, there could be no such thing as a vacuum.

Torricelli's mathematical scientific pursuits were cut short by his sudden death from typhoid in 1647 at age thirty-nine. Other scientists and philosophers further removed from the Roman Inquisition—including Irishman Robert Boyle, a cofounder of the Royal Society of London, and Frenchman Blaise Pascal—extended and refined Torricelli's experiments. By 1670, barometers were being made as weather instruments for the noble rich. By the late 1800s, the barometer was used to measure not only air pressure, but altitude as well. Its popularity as an instrument of technology was akin to that of the computer today. The value of the barometric principles Torricelli discovered cannot be overstated.

4

Death Comes to Eugene

In the fall of 1962, it was pretty common for Larry and JoAnn Johnson to be joined by their downstairs neighbors, Sam and Walli Grubb, for Friday night socials featuring spaghetti, cheap red wine, and a game or two of Monopoly. Such was the austere life of cash-strapped couples living in "Fertility Flats," the nickname married students had bestowed on the drafty, converted army barracks officially known as the Amazon Housing Project, a few blocks southwest of the University of Oregon campus in Eugene. As the nickname for the housing project suggested, many of the hundred or so couples living there had young kids, infants, or a child on the way: the Grubbs had a two-year-old son, and JoAnn Johnson was eight months pregnant. Larry Johnson, a stocky guy who loved to hunt and fish, was a senior majoring in architecture. He was just six days shy of his twenty-second birthday.

When the Grubbs went upstairs about 5 p.m., October 12, the Columbus Day Storm had already reached Eugene. But the couples weren't troubled by the turbulent weather: they didn't intend to venture out that night. They felt safe and secure in the two-story apartment complex they called home. Like most Pacific Northwest residents, they didn't know the details of the storm. They knew nothing about the turbulent mass of angry black clouds that reached over Oregon's coastal mountain range and swept up the Willamette Valley into the heavily forested college town of Eugene late Friday afternoon. Just before 4 p.m. at the Eugene airport, the temperature shot up from 50 degrees Fahrenheit to 61 degrees, and then the winds began to howl, including gusts up to 86 miles per hour at 3:50 p.m., a hurricane-force velocity never before experienced, or at least never before recorded, in the most populated city in the south Willamette Valley of Oregon. With most Pacific Northwest windstorms, the strongest winds develop ahead of the storm front and falling barometric pressure. The Columbus Day Storm behaved differently: the strongest winds were in areas of rising pressure and warm air that followed on the heels of the cold front.

The couples did realize it was more than a typical fall windstorm when a section of window blew apart in the Johnson's cramped bedroom. Johnson and Grubb grabbed a piece of cardboard to cover the window to keep rain from soaking the bed, which sat just two feet from the window. This shouldn't take too long, the college buddies thought. It was just a little patch job.

The worst windstorm to ever strike the West Coast in modern history was ripping through Eugene. One block to the south of the married student apartments, winds tore the pitched roof off Roosevelt Junior High School, sending it sailing on a lethal trajectory right at the Johnson's south-facing bedroom. "In my mind it was like slow motion, but it happened too fast for us to escape," Grubb, a retired contractor living in San Felipe, Baja, Mexico, said. "We tried to turn and duck, but it was too late. It was all one big explosion when the roofing hit the apartment wall."

Johnson took the brunt of the powerful blow, perhaps saving Grubb's life. Johnson landed on top of Grubb, his head dug into Grubb's chest, breaking two of his ribs. Grubb also suffered several head lacerations and a broken left arm. Johnson fared much worse: a jagged piece of wood pierced his chest. Both men were knocked unconscious. Neighbors rushed to the scene and dragged the two bloody, gravely wounded men to the living room. Grubb, a premed student just months removed from a two-year stint as a navy hospital corpsman, regained consciousness and began CPR on Johnson, who was still alive. With help from a neighbor, they had Johnson breathing again, Grubb recalled. That's when responders from the Eugene Fire Department arrived and took over first aid. Weak from loss of blood, Grubb passed out again.

Hours later, Grubb woke up at Sacred Heart Hospital in Eugene and learned that Johnson was dead. JoAnn Johnson was admitted to the hospital and treated for shock. Somehow, Grubb convinced hospital staff to let him go home to his wife and young son, who he helped settle in to bed. As Grubb undressed, he backed into a free-standing oil heater in the living room of their relatively unscathed apartment and burned his rear end. "My whole head was wrapped in a big turban, my left arm was in a cast, and my ribs were throbbing," he recalled. "At this point the burned butt was just comical."

The night's drama wasn't over yet: Grubb thought he heard someone breaking into his apartment. He stood by the door, preparing to whack the intruder with his cast. It turned out to be his father, who had rushed down to Eugene from Portland after learning of the horrific accident involving his son. The father and son headed to downtown Eugene and found an open bar for a drink before calling it a night.

Johnson was one of five storm fatalities in Eugene that day. Terror and confusion reigned on the city streets and in the neighborhoods along the banks of the Willamette and MacKenzie Rivers. No other community on the path of the mighty windstorm suffered greater human loss.

South of the Amazon Housing Project, Curtis J. Ray, eighty-four, was in his wife's room at Sunset Home, a large retirement center and the aging couple's home for the past two years. He stood by her bed in the gathering darkness, trying to comfort her from the fury of the wind that buffeted the nursing home. Born in Chicago, Illinois, in 1877, Ray was just eleven days shy of his eighty-fifth birthday. He was a master candy maker who had learned the craft as a young boy. He and his wife, Martha, moved to Eugene in the 1930s, raised a daughter and had been blessed with a grandson.

Martha Ray's room sat next to the Sunset Home chapel, a brick building with a wood and glass front. The chapel was a barn-like structure that measured seventy-five feet long, thirty feet wide and twenty feet tall. The wind clawed at the chapel and attacked the façade, which, according to Sunset Home superintendent W. J. Tweet, "fairly exploded" in an outward burst of glass and splintered wood. The chapel's south wall crumbled next. Then the roof collapsed and the remaining walls toppled. An avalanche of bricks plunged through the ceiling of Ray's room, burying Curtis Ray and his wife beneath a pile of debris. He was crushed to death. She escaped with a fractured wrist and leg injuries, one of forty-five Eugene residents injured in the storm.

Three other apparent heart attack victims added to the college town's death toll. Margaret Hilda Blythe, seventy-three, died of an apparent heart attack in her Thirty-Third Avenue home not far from Amazon Park. Blythe, a native of Middlesex, England, called Eugene police for help after the front window in her home blew apart during the storm. The dispatcher told her police were swamped with calls: she would have to wait for assistance. The dispatcher warned her to stay away from the window, hung up, and answered the next storm-related call. Not long after, Blythe died in her home. Police theorized that she was upset by the damage to her home, which triggered a fatal heart attack.

Herbert Harold Beeson, sixty-five, was found shortly after the storm struck, lying on a sidewalk on Tenth Avenue in downtown Eugene, a few steps from the laundry where he worked, another victim of an apparent heart attack. A veteran of World War I, he was a charter member of the Eugene Yacht Club and had worked for many years in the dry cleaning business in Eugene. DeWitt O. Bevans, fifty-four, suffered a heart attack and died Friday night attempting to repair storm damage to the roof of his Ascot Drive home. Bevans was the Eugene district manager for Safeway Stores, Inc., and had worked for the grocery chain since 1931.

Five fatalities seem like a cruel number in a community of fifty-three thousand, just five days shy of its centennial birthday celebration. Portland, the state's largest city with a population 373,000 at the time, also in the direct path of the storm, experienced four storm deaths.

But the death toll in Eugene could have been so much worse. University of Oregon English instructor Lucile Payne was preparing to speak to nine hundred high school journalism students gathered on campus for a Saturday conference. Those plans were waylaid when she was caught downtown Friday as the windstorm struck. She ducked into a Sears and Roebuck store for safety. She joined two other women, a store clerk and a customer, in front of a big storefront window looking out on Tenth Street just in time to watch the roof blow off the old city hall building. Just then the window imploded, propelling shards of glass through the air. Payne turned and ran, the wind pushing at her back. "Pieces of glass were streaming past me," the short-story writer and teacher said in an account she wrote for the Eugene *Register-Guard* newspaper. "One piece took a neat little slice out of my jaw, but I didn't even feel it. The only thing I really felt was surprise. I was quite certain I was going to be killed, or maybe had been already but didn't know it yet."

The other two onlookers were also cut by flying glass. The store clerk was hospitalized with deep lacerations to her legs, and the customer credited a big black hat she was wearing with deflecting jagged chunks of glass, perhaps saving her life. "I don't think I'll ever stand in front of a window again, even on a sunny, clear day without a breeze," Payne wrote a few days after the storm. "I plan to do quite a lot of cowering from now on."

About a mile southeast of downtown, the University of Oregon campus was the scene of risky behavior as college students stood around in clusters, watching trees that were as old as, or older than, their heavily wooded, eighty-five-year-old campus crash to the ground. A crowd of seventy-five gathered in front of Allen Hall to witness the final moments of two stately fir trees. Suddenly a strong gust blew down eight trees around them, including two that fell right behind the crowd. Observers from safer vantage points in nearby campus buildings watched in horror and screamed warnings at the crowd. The students came to their senses and bolted for cover.

The storm was responsible for at least twenty-seven direct and indirect deaths in Oregon. The victims ranged in age from the eighty-four to two years old. Falling trees and limbs were responsible for ten of the storm-related deaths in Oregon. Heart attacks claimed another nine. Three, including Ray, were killed by collapsed buildings. Two storm victims were electrocuted by downed power lines. Two others, including Johnson, were struck by windblown objects, and Fred Shirley, eighty-three, of Salem, died Saturday after falling from his roof while making storm repairs.

The death toll from the Columbus Day Storm is an inexact number, as low as forty-six and as high as sixty-five. The higher number rings true if the Columbus Day Storm is viewed as the main attraction, bookended by associated West Coast storms Thursday and Saturday. The three-day blast of stormy weather between October 11 and 13 resulted in a record rainfall of seven to twenty inches in the San Francisco–Oakland area. The three-day death toll in the Bay Area climbed to nineteen, including three people buried in mudslides, one electrocuted, one drowned at sea, and one killed by a falling tree. The rest died in storm-related car accidents. The eleven lives lost in Washington State included seven storm victims struck by falling trees, two who suffered heart attacks, one who drowned, and another who fell off a roof in Seattle, trying to remount a television antenna. The Columbus Day Storm also claimed eight lives in and around Vancouver, British Columbia, and the less powerful Thursday predecessor took another life. The deaths followed an all-too-familiar pattern: three from falling trees, two from heart attacks, one from a car crash, one an electrocution, and one a drowning.

In the aftermath of the deadly storm, some family members couldn't agree if a death in the family was caused by the storm or not. Take the case of Burnelle White, a state parks employee in the coastal town of Newport, Oregon. He died of a heart attack Monday, October 15, while cleaning up storm debris at Yaquina Bay State Park. "I lost my dad to the storm," a daughter, Daphne Lawrence, said with conviction. Another daughter, Jennifer White, disagreed, noting her father's preexisting heart condition. "I don't think he was a fatality of the storm," White said.

White doesn't show up as a storm victim in any official or unofficial tally that I've found. Neither did Vancouver, British Columbia, city road chief W. M. Robinson, who died of a heart attack at the wheel of his car Monday, October 15, after working nonstop over the weekend directing cleanup of fallen trees and other storm debris from city streets. However, at least three Salem men who suffered fatal heart attacks cleaning up storm debris October 13–14 appear as storm victims in newspaper stories about the storm death toll.

Determining the death count from a natural disaster is not an exact science. For example, Clifford R. Brown, forty-five, collapsed and died on his bedroom floor during a fire in his Olympia, Washington–area home, October 12, 1962. The headline in the *Daily Olympian* newspaper read, "Man Dies in Fire as Storm Hits Area." The story suggests that flames from the fire swept through the home quickly, propelled by the stormy winds. Brown was not considered a storm fatality by most accounts. Nor was Fredrick Johnson, forty-eight, a patient at Oregon State Hospital in Salem, who suffered a heart attack in bed after first being caught outside in the raging winds that raced through the mental hospital Friday night.

The great force multipliers for the Columbus Day Storm were the mighty trees of the Pacific Northwest: Douglas-fir, oak, hemlock, cedar, maple, and more. They suffered their share of damage: whole trees uprooted, treetops snapped off, trunks splintered, and heavy branches flung through the air. Falling trees and tree limbs get credit for the most storm deaths—twenty-one. In most cases they came crashing down on people in their homes or in their vehicles traveling on country roads and city streets.

The next deadliest toll—fourteen—was linked to car accidents, mostly in Northern California, where the driving wind and rain turned the roads treacherous. A close third, at thirteen, were heart attacks, both during the

Cabins at a resort on the road to Mount St. Helens in southwest Washington were hard hit by falling trees. Courtesy of the Cowlitz County Historical Museum, Kelso, Washington.

direct storm assault and in the day or two after the storm. It comes as no surprise that heart attacks reared their ugly heads in and after the storm. Modern-day cardiology was in its infancy—bypass grafting of clogged arteries, angioplasty procedures to open up clogged arteries, and today's array of heart-disease-fighting medicines were still years away. Back then, when doctors found a blockage in an artery, there was no good way to treat it, said Dr. Christopher Wolfe, an Olympia, Washington–based cardiologist.

In addition, more than 42 percent of the adult US population smoked cigarettes in 1964, the same year US Surgeon General Luther Terry issued a landmark public health study showing direct links between cigarette smoking and lung cancer, and likely links between smoking and heart disease, too. The health threat, combined with subsequent bans on cigarette advertising and on smoking in workplaces and public settings, as well as requiring warning labels on cigarette packages, helped reduce smoking rates to 19 percent of the US population by 2014.

Falling trees, car crashes, and heart attacks were the big three causes of death. Everything else—electrocutions, collapsed buildings, drownings, flying debris, mudslides, and tumbles from roofs—measured two or three fatalities each. Oregon's Willamette Valley was truly ground zero for the Columbus Day Storm. It was home to twenty-seven deaths, followed by nineteen in California, eleven in Washington, and eight in British Columbia.

Trauma from the fatal accident, combined with classes he missed while recovering from his own injuries, knocked Grubb out of the University of Oregon premed program. He dropped out of college, moved his young family to Portland, and took a job as a traveling salesman, selling plastic pipe fittings. He eventually completed college at Portland State University and worked much of his adult life as a contractor in Tacoma, Washington. Later in his career he served as a construction project consultant and expert witness in construction lawsuits. Divorced twice and married three times, he lost track of his college friend's widow. He spends most of his time south of the border, living what he calls a profligate life, taking it one day at a time, his sunrises with coffee and sunsets with gin. "It was a life-changing event," Grubb said of the gruesome storm accident.

JoAnn Johnson moved back to her parents' home in Sacramento and gave birth to a healthy baby boy she named Larry Johnson Jr. She also hired a young Eugene, Oregon, attorney, Art Johnson—no relation—to investigate the fatal accident. Johnson, a Harvard Law School graduate and Air Force veteran of the Korean War, filed a lawsuit in Lane County Circuit Court, alleging that the central section of the junior high school roof that lifted in the wind and crashed into the apartment had not been properly attached to the schoolbuilding walls. He sued the school district, school architect, and contractor for the maximum Oregon State law allowed at the time in a wrongful death lawsuit—$25,000. Circuit Judge Edward Leavy signed a motion to dismiss the case in December 1964 after the parties reached an out-of-court settlement that dropped the school district from the case, according to Art Johnson, who, in the spring of 2014 was eighty and still active in the Eugene family law practice. Grubb received a check for $1,000 from the settlement. "I've often thought that I should have pursued the case, because there was no liability limit for an injury," he said. "I could have made much more and given it to the widow."

One bizarre footnote to the deadly storm in Eugene: Jimmy Gillette, the first Eugene Fire Department employee to arrive at the Johnson's apartment that Friday night, died a violent death fifty years later at his south Eugene home. Gillette, seventy-three, and his partner, retired University of Oregon music school dean Anne Dhu McLucas, were bludgeoned to death September 7, 2012, by Gillette's wrench-wielding son, Johan Gillette. The son was convicted in March 2014 of two counts of aggravated murder and sentenced to life in prison.

5
Coastal Chaos

The US Coast Guard and its predecessor, the US Life Saving Service, has had a presence in Yaquina Bay since 1896, serving as a guardian of the Central Oregon coastline, helping ships in distress and patrolling hundreds of miles of beaches. The men and women stationed there in the gritty, weather-beaten coastal town of Newport, home to the state's largest fishing fleet and scenic playground for ocean tourists since the 1860s, are no strangers to stormy weather. When the winds howl and the rain slants sideways, that's when they are their busiest. But nothing in recent memory quite like the Columbus Day Storm of October 12, 1962, had paid a visit to Yaquina Bay and Newport.

At 4 p.m., Coast Guard seaman William Fuhlrodt was on duty in the Coast Guard's lookout station at Yaquina Bay when the storm descended on his outpost with sudden fury. He stared in disbelief at the wind gauge as the wind gusted and the needle pushed to the far right, bumping up against the restraining peg. The anemometer was designed to register winds only up to 120 miles per hour. This wind was mightier than the wind gauge could measure.

Three times in an hour the needle reached a maximum reading. At 5 p.m., Chief Bosun's Mate Giles Vanderhoof had seen enough. Fearing for his crew's life, he ordered the lookout station evacuated. Fuhlrodt didn't hesitate to obey the order. He clambered down the ladder as quick as he could, his feet barely touching the rings as he descended. Vanderhoof later estimated the peak wind gusts screamed into the bay at 140 miles per hour.

Newport, a town of fishermen, artisans, California refugees, and merchants, grew up along and near the serpent-head shape of the Yaquina Bay shoreline. The early pioneers in the mid-nineteenth century were not deterred by the isolation, the dreary weather, or the words of Cyrus Olney, an Oregon

Territory lawyer and politician who was drawn to the area by an 1852 ship-wreck. He had this to say about Newport's prospects as a place to settle and develop: "In time it may become the most available avenue to and from the Upper Willamette Valley; but the first settlers must suffer many inconve-niences and make great expenditures and rely on an indefinite future."

Early on, the sumptuous Yaquina Bay oysters, craved in San Francisco, and burgeoning tourism were the growth industries for this remote coastal outpost. Newport pioneer, postmaster, and bay front land developer Sam Case, and his associate Dr. James R. Bayley, in 1866 constructed the first resort hotel on the Oregon coast—Ocean House Hotel. It sat on a small terrace and straddled the bay and ocean, sharing a vantage point with the Yaquina Bay Lighthouse. Built in 1871, the lighthouse was a short-lived navigation aid for the busiest port between San Francisco and Puget Sound. It was replaced in 1874 by the lighthouse perched north of town at Yaquina Head, a headland sculpted by jumbled heaps of basalt that speak to volcanic eruptions and geologic upheaval millions of years ago.

Newport's first foothold, now the historic Bayfront district, which began life as a muddy, southeast-facing implant of a few buildings—a mer-cantile, saloons, hotels, and, yes, a brewery—has morphed a hundred years later into a nine-block-long string of restaurants, bars, chowder houses, fish-processing plants, and curio shops, as well as the home of the first Rogue Brewery—which suggests beer-making belongs on Yaquina Bay then, now, and forever. And what brewpub, what marina, what bay-bound community wouldn't like to be dissected by the graceful green arches and art deco spires of the grandest of all the Oregon coastal bridges? To most, it's the Yaquina Bay Bridge, one of five coastal bridges completed in 1936, a Franklin Delano Roosevelt New Deal accomplishment designed by Conde McCullough, a talented bridge engineer who was not afraid to spice up his work with a hint of romance.

I pieced together my impression of Newport on a calm October day in 2013, soaking up the image of the bridge as I continued north past the modern-day resorts perched on ocean-side cliffs before they yield to Newport's gentrified Nye Beach District, where the gray-brown shingle-sided cafes, gift shops, coffeehouse, bookstores, and literary-themed Sylvia Beach Hotel cling to hilly streets that offer either exposure to, or protec-tion from, the ever-present ocean breezes. Newport weather on this early fall day was a far cry from those 250 hours a year when the wind blows more than 50 miles per hour. The temperature was in the fifties, where it resides much of the year. The rain rested in a temporary reprieve from what approaches a ninety-inch-a-year assault.

This day, all was well on this windswept patch of land that is home to ten thousand occupants shoehorned between the relentless sea to the west and a jagged-toothed, coastal mountain range to the east. The light winds seemed to say, "Enjoy it while you can, this calm before the next inevitable, damaging storm." The really big blows born of remnant, reenergized typhoons from the western Pacific Ocean make landfall on the Central Oregon coast every fifteen to twenty years, not like clockwork, but inevitable enough to keep coastal residents wary. These sudden blasts of wind and rain are slightly more frequent than damaging Pacific Northwest earthquakes, and just as frightening.

Newport old-timers are no strangers to windy weather that can bring gusts of 100 miles per hour or more sweeping over and through their homes and businesses. But one windstorm was etched in their memories like no other. That would be the storm of the century, the Columbus Day Storm of 1962. "I've lived in Newport since 1941, and I can't remember a storm stronger than that," former Newport fire chief Don Rowley said on the eve of the fifty-first anniversary of the Columbus Day Storm.

Rowley, a volunteer firefighter in 1962, was busy that mid-October night. He helped families with torn rooftops and blown-out windows patch their homes or find emergency shelter. He cleared city streets of fallen trees, downed power lines, scattered billboards, and other wind-blown debris. However, his hard work and goodwill did not provide personal immunity from the storm's wrath. He experienced his own double whammy of storm damage. The winds blew down the south wall of his Newport gas station. Late that night—and finally off duty—he had to call for help from his fellow firefighters to cut up and remove fourteen fallen trees strewn across the driveway to his home, which lost a dormer to the gusting winds. Don Davis, who arrived in town on October 1, 1962, to start his new job as Newport city manager, echoed Rowley's sentiments. "I can think of about a half-dozen storms that got my attention over the years. But the Columbus Day Storm is the granddaddy of them all."

The storm earned its stature in a variety of ways. Newport was punched with wind gusts estimated at 138 miles per hour, perhaps the fourth-highest along the path of the storm, and the equivalent of an extreme category 4 hurricane. The uncertainty on peak winds is easy to explain: at many locations, wind gauges were not calibrated to measure the strength of this storm's wind, or they were ripped from their stations by the wind's force, leaving weather observers to take educated guesses at wind velocities. Bayfront, which confronted the wind head-on along the north side of Yaquina Bay, was enveloped in what was reported to be an astonishing 150-foot-high wall of saltwater spray. The wind plastered seaweed on the windows of nearby homes and the bay took on the look of boiling water. Old wooden waterfront warehouses and the crudely named "stink plant" (think waste from fish processing) were blown apart by the winds, scattering woody debris in the bay's roiled waters.

The winds tore off the storefront of Mo Niemi's original Mo's clam chowder and seafood restaurant on Bayfront, dropping it on a pickup truck parked on Bay Boulevard. Her son, Sonny Niemi, gathered the wreckage and dumped it in the bay. The hurricane-force winds blew the front face off the Walker Building, then home to the Elks Lodge and a bowling alley, and propelled it four feet onto Hurbert Street, crushing a parked car. Rebuilt after the storm, the two-story wood and brick structure housed a barbershop, dance studio, pub, and Pacific Coast Plumbing, when I visited. "You

Winds whisked away the Walker Building storefront in downtown Newport, Oregon, at the peak of the storm. Courtesy of the Lincoln County Historical Society, Newport, Oregon.

Yaquina View Elementary School in Newport, Oregon, was no match for the Columbus Day Storm. Courtesy of the Lincoln County Historical Society, Newport, Oregon.

can see why the building took a beating in the storm," plumbing business co-owner Bill Jernigan said on the eve of the storm's anniversary. "It's the highest spot in town."

The widespread damage in Newport, estimated at more than $39 million, prompted Governor Mark Hatfield to declare the town a disaster area. About thirty National Guardsmen joined state and city police to patrol the town and protect against looters Friday night after the storm knocked out electricity and pitched most of Lincoln County into darkness. By noon Saturday, power was restored in Newport, but a 10 p.m. curfew was imposed to protect against looting and vandals. Schools throughout Newport took a real beating, especially the Yaquina View Elementary School, which was then just two years old, perched on an exposed bluff overlooking the bay. A wall and roof of the school's six-room south side were dismantled by the wind. The school, which suffered more than $1 million damage, looked like the victim of a terrorist's bombing, with splintered glass and wood lying about in the storm's aftermath. The roofs of four other schools in the Newport area were severely damaged.

The winds and falling trees destroyed twenty homes in Newport and damaged dozens of others. No one was killed or seriously injured in the Newport area, but there were numerous close calls. The winds pushed a

Nye Beach rental home owned by Bob Whisler off its foundation, sending it sliding toward the ocean cliff and pounding surf seventy-five feet below. A family caught by surprise in the crippled home tried to exit through the doors, but the doors were knocked askew and jammed shut. The family escaped out a back window, fearful the home might plunge over the cliff. It didn't, but it took Whisler eighteen months to repair the damage to his Newport rental properties.

Six miles east of Newport, in the Yaquina River town of Toledo, Ray Wilkinson was giving a music lesson Friday afternoon when he heard word of the fast-approaching storm. He headed for his Depoe Bay home on the coast thirteen miles north of Newport, trying to get a jump on the storm, but the storm jumped him instead. First, a big conifer tree fell across Highway 101 directly in front of his car at Otter Crest near his home, blocking the road. He tried to turn around, but a second tree fell behind him. Then a third tree fell on his car while he was in it. He escaped shaken, but unharmed.

A big storm that had slammed the Oregon coast on Thursday had already forced the Newport fishing fleet back to the docks when the more ferocious storm hit Friday. Most of the 150 vessels either lashed themselves doubly tight to the docks or motored upriver six miles to Toledo to seek more sheltered waters, recalled Ray Hall, a Newport fisherman who started crabbing, shrimping, and bottom-fishing in 1932. He remembered a few boats blown aground in the bay, and in the river, but the reports in the newspapers at the time reported only one grounded fishing vessel and one sunken rowboat in the bay.

Hall told me he retired from fishing three years after the storm, but not before he helped supply crab for Newport's infamous May Day crab festivals. The 1938 inaugural event, designed to draw attention to Newport and boost markets for Dungeness crab, drew twenty-five thousand people to a town of two thousand for a free crab lunch. "After the banquet, the street department employed a dump truck and two men for three days to sweep up the cracked crab shells," Newport historian Steve Wyatt recounted in his 1999 collection of Newport stories titled *The Bayfront Book*. Eventually crab grew too expensive to hand out for free, and Newport's crab festival faded away in 1951.

Hall knew of the storm's approach from listening to a Coos Bay radio station reporting hurricane-force winds fast-approaching from the south.

Burnelle and Teddy White were also in the know. The husband and wife team were Newport-based weather observers employed by the United States Weather Bureau. They were charged with taking several weather readings daily from a weather station mounted behind their Newport home. They phoned them to the Salem weather bureau office, where they were used to shape weather forecasts.

Some twenty-five years after the storm, Teddy Smith, formerly Teddy White, wrote a two-page memoir of the storm and the events that followed. "On Friday, October 12, I realized that there was a strong wind coming up, not unusual for the coast. Thinking that it was probably cool outside, I put on a warm jacket and went out to read my instruments. The wind was very strong, but much to my surprise, it was very warm." The other thing that caught her eye was a 4 p.m. barometric pressure reading of 28.50, the lowest reading they had ever recorded. The winds were gusting at 80 miles an hour. She and two of her children jumped in their car and headed for Yaquina Bay State Park, where her husband worked on the parks maintenance crew.

As the three drove through uptown Newport they saw storefront windows and street lights blown apart, daughter Jennifer White recalled. The family arrived at the state park on the bluff above the bay and waited for Burnelle. Rocks from the parking lot pelted the car, which was rocking in the wind as the father scrambled into the car. The family drove home and huddled inside while the storm raked Newport. Chunks of the neighbor's tarpaper roof bounced off the side of their house, barely missing the windows. Nobody in the White family slept much that night; the winds were too noisy and frightening. In the morning, they surveyed the neighborhood for damage. They fared well, but an uprooted tree hung dangerously over a neighbor's home. It was cut down before it fell.

Monday, October 15, was the White's thirtieth wedding anniversary. Tired from all the weekend cleanup and lack of sleep, Burnelle went to work, despite a stomachache and nagging pain in his chest. The state park was a mess of downed trees and limbs and posed a danger to all the people who showed up at the park to witness the storm damage. In the middle of the afternoon, the crew boss showed up at the family door and told Teddy White that Burnelle had fallen on a park path near the Yaquina Bay Lighthouse and had been taken to the hospital. He died of a heart attack before his wife could reach his side. "I wasn't alone, as there were other incidents of death and accidents caused by the storm," she said. "Those stories belong to other people."

My October afternoon visit to Nye Beach coincided with a Newport High School cross-country meet—six boy's and girl's teams sharing a three-kilometer oblong course laid out north–south on the surf-touched sands of Nye Beach. Nye Beach was no place to run, or beachcomb, or enjoy in the era of the Columbus Day Storm. The beach was fouled by thick, noxious foam generated by a Georgia Pacific Corp. pulp mill in nearby Toledo, Oregon. The mill effluent was piped to Nye Beach and discharged a scant 1,075 feet from high tide. Outraged Newport citizens, who watched as tourists fled their town, petitioned the state for an end to the dumping. The best they got in 1964 was a nearly four-thousand-foot extension of the discharge, which is still in place today.

A northerly breeze both aided and buffeted the youthful runners as they trudged along, seeking purchase in the shifting sands licked by the salty, probing tongue of an incoming tide. Overhead, gulls and brown pelicans plowed headlong, stoic and determined, into the wind above the pounding, frothy surf. I conjured up an image of former University of Oregon and Olympic long-distance runner Steve Prefontaine watching the youthful runners carry on his joy-of-running legacy. "Pre," a child of the sixties who grew up setting state and national high school track records at another Oregon coastal town—Coos Bay—before he crashed his car in Eugene and died in 1975, would surely smile if he could see these teens running in his footsteps.

An ocean beach playing host to a high school track meet was made possible more than one hundred years ago by a far-sighted politician—Governor Oswald West—who crafted and cajoled through the state legislature a bill making the state's 362 miles of ocean beaches a public highway—not for sale. Free, unimpeded public use of the beaches was reaffirmed when Governor Tom McCall signed the Beach Bill in 1967, and the public ownership continues to this day, arguably Oregon's greatest claim to fame.

It was my last night in Newport. The sun dipped into the ocean as I walked the high-tide line of Agate Beach on the northern outskirts of town. Waves pounded the Yaquina headland, spewing into the air a spray that formed a gauzy haze. Heaps of bull kelp the size of sofas dotted the beach, waiting to be sucked back to sea by a forceful high tide. The seasonal sand dunes were scalloped and sculpted. Bits of dune grass maintained a tenuous grip on the beach, soon to be uprooted by the force and fury of fall and winter storms.

I looked closer at the wrack line and wished I hadn't: specks of plastic, colorful but depressing, formed a red, white, green, blue, and pink display—all the colors of a manmade plot to pollute. Adding to the dismal array of plastic were the carcasses of a red-tailed hawk, a pigeon guillemot, and a foul-smelling deer. Two hours later, the antidote for those depressing images on the beach emerged in the sky in the form of a seahorse-shaped cloud cantering across a swollen crescent moon. The specter of stormy weather on the horizon, locked into the October calendar, became a bit easier to accept.

On a clear day, the views from the old Mount Hebo Air Force Station, perched high in the northern Oregon Coast Range, can stun the senses. To the west, the true horizon—that clean, clear line between the Pacific Ocean and the sky—comes into view. To the east, five peaks in the Cascade Range reveal their snowy tops—Mount Hood, Mount Jefferson, Mount St. Helens, Mount Adams, and even the mightiest of them all, the 14,411-foot-tall Mount Rainier, some 160 miles away. To the north and south, coastal bays and seaside villages sit smeared by the misty haze from surf breaking on the shore and crashing on the jagged headlands and cliffs that unfold for miles in each direction. The view stretches well beyond Cape Meares and Tillamook Bay to the north and past Neskowin Beach State Park to the south. These picture-perfect days are elusive, rarer than the endangered silverspot butterflies that call the mountain's meadows home.

Carol Johnson, a former public affairs officer stationed at the Hebo Ranger District of the Siuslaw National Forest from 1992 to 2003, has lived in the shadow of Mount Hebo since 1966. She has visited the mountain for work, or to picnic, hike, and pick wild strawberries, more times than she can count. "I've lived here fifty years and only seen the full panoramic view six times," she said.

Rising 3,175 feet from the dense stands of Douglas-fir, hemlock, and spruce below, the exposed tabletop that is the peak of Mount Hebo was dubbed "Heave Ho" by Oregon pioneers. It's a fitting way to describe the way the mountain seems to erupt from the surrounding terrain. The mountain was traversed often, first by Native Americans, then by mid-nineteenth-century pioneers. The mountain straddles a path that linked the Tillamook Valley on the Oregon coast to the interior Willamette Valley, the only such route until white settlers completed a wagon road through the coastal mountains in 1882. With the passing years, Heave Ho morphed into Hevo, then Hebo.

In the late 1950s, the US military-industrial complex combined to build an air defense system known as the Semi-Automatic Ground Environment, which was the largest and fastest computer system assembled at the dawn of the computer age. The system linked 140 military radar stations to twenty-four "directions centers" and three "combat centers" scattered across the country. The massive computer mainframes built by IBM, each covering a half acre of floor space, received digital data from the radar stations via AT&T rotary phones. The computers deciphered the data to determine whether an aircraft was friend or foe. The military used the computer system, valued at $67 billion in today's dollars, to decide whether or not to activate missiles and fighter planes to intercept Russian bombers before they could drop their bombs on American cities and military bases. The Cold War was raging, and nuclear winter was just a lapse in judgment, a misread bit of data, a button-push away.

Mount Hebo was a logical choice for one of those radar stations. The air force secured sixty-seven acres from the United States Forest Service and built in 1957 what amounted to a small compound on top of Mount Hebo, including a radar and communications operations center, barracks, twenty-seven units of family housing, a dining hall, power plant, gym, motor pool, and helicopter pad. That Soviet threat, as real as it was in the Cold War, never materialized in the shape of an advancing bomber. The weather was the one enemy that relentlessly punished the psyche of those assigned to Mount Hebo.

On October 11, 1962, just days away from the showdown with the Soviets in Cuba, Russian bombers armed with atomic bombs were the least of worries for members of the 689th Aircraft Control and Warning Squadron. Their radar station was under assault by hurricane- force winds that were a precursor to the Columbus Day Storm. Two straight days of winds that topped 100 miles per hour forced evacuation of the radar station and sent personnel scurrying to the barracks for protection.

One of the tallest mountains in the Coast Range, Mount Hebo is fully exposed to storms sweeping off the Pacific Ocean from the south-southwest. The 160 or so men and women stationed at Mount Hebo Air Force Station at any given time between 1957 and 1979 experienced some of the most severe winter weather conditions of any military base in the lower forty-eight states, said the base's first radar maintenance officer, retired air force captain Dave E. Casteel. "The bad thing about Mount Hebo was the southwest-facing valley that reached from the ocean all the way to the top of the mountain," said Robert Tillmon, a retired long-haul truck driver from Houston, Texas, who was an airman second class stationed there from May of 1962 to September 1964. "The valley just funneled the wind right into the air force station."

It didn't take long, upon his arrival at Mount Hebo, for Tillmon to realize bad weather would be a constant companion. "Here it was May and the first thing they issued me was Arctic weather gear," the former radar maintenance technician said. Rain and snow piled up at the top of the mountain at the rate of 180 inches a year, burying buildings until the spring thaws. Fall and winter storms routinely delivered winds that topped 50 miles per hour with gusts of more than 100 miles per hour. The metal buildings that dotted the base were tied down with cables to keep from blowing away. Casteel said the standing joke among the men stationed there went something like this: The fence around the high-security radar installation wasn't built to keep out trespassers—it was designed to keep people from blowing off the mountain.

The base featured a network of corrugated metal tunnels that served as safe passageways for servicemen who otherwise would be exposed to the hellish winds, sleet, and snowstorms that struck suddenly in the fall and winter months. Icicles that morphed into lethal projectiles were another reason for the construction of the metal tunnels, Casteel said. Ice formed on the tips of the radar antennae that rotated at 23 miles per hour more than one hundred feet above ground level. In the winter of 1961, base commander Major Raymond R. Robinson observed jagged chunks of ice sailing through the air and striking the ground. He asked Casteel if he should order hard hats for the operations personnel. Casteel made some hurried calculations and concluded that when the radar was rotating, the ice would hit about fifty feet out from the radar tower walls and would be traveling 100 miles per hour when it hit the ground. He told the base commander that he doubted that hard hats would be much help. Base officials opted for the metal tunnels instead.

The double-whammy storm in 1962 arrived at Mount Hebo AFS during construction of one of the largest dome-shielded radar systems in the world. The dome consisted of 1,376 hexagon-shaped fiberglass pieces that weighed from 90 pounds to 150 pounds, said Mark Cole, who was stationed at Mount Hebo on October 12, 1962. Once fastened together, the panels would make a radar shield 100 feet tall and 145 feet in diameter. The radome was set for completion by November 1, just in advance of the severe storm season. Or so the project construction team thought. The radome was about 54 percent complete when the Columbus Day Storm struck, Cole said. As the winds engulfed the mountaintop, the wind gauges at the base pegged out at 131 miles per hour, surpassing Thursday's reading of 120 miles per hour.

It was estimated that the Friday winds gusted at up to 170 miles per hour, which is a wind speed associated with a category 5, catastrophic hurricane. It was the second-highest wind velocity estimated for the storm, second only to the 179 miles per hour calculated at the US Coast Guard station at Cape Blanco along the Southern Oregon coast. To put the Mount Hebo blow in perspective, the wind gusts there rivalled the three windiest category 5 hurricanes to make landfall in the lower forty-eight states. Those were two Gulf of Mexico monsters—Hurricane Camille in 1969 (259 deaths) and Hurricane Andrew in 1992 (65 direct and indirect deaths) and an unnamed, 1935 Labor Day storm that killed 408 in the Florida Keys.

"We knew a pretty bad storm was coming, but no one knew it would be that bad," Tillmon said. Tillmon holed up in the base barracks, a cinder block building with small windows about one-half mile from the radar installation. As their world seemed to be on the verge of blowing apart, Tillmon and other off-duty personnel were joined by fifty members of the squadron who were evacuated from the radar operations station. Like a home intruder in the night, the howling winds found the opening at the top of the partially built radar shield. The wind blew the radar cover apart from the inside out. The panels, several inches thick, flew through the air like Frisbees. The winds ripped the antennae off the operating radar and knocked a smaller, adjoining radar station offline. Some sections of radar shielding flew off the side of the mountain and sliced into Douglas-fir and

Winds from the fierce Columbus Day Storm tore apart the radar at the Mount Hebo Air Force Station. Courtesy of David Casteel.

hemlock trees in the valley below. Cole recalled peering out the window and watching cars in the parking lot bouncing up and down in the wind, their shock absorbers groaning. Miraculously, no one was injured.

At daybreak Saturday, station personnel surveyed the damage. Cole described a scene akin to a junkyard, with twisted scaffolding, metal siding, fiberglass, and other construction debris scattered on the ground and hanging precariously over the radar operations building. "It looked like a bomb had gone off," Tillmon said. Crews scrambled to clean up the mess and bring the radar base back online within a few days, Casteel said. Construction of the large radome was completed in 1963. The structure succumbed to high winds and lightning again in January 1964. A third radome of a slightly different design was destroyed in a 1968 windstorm.

While Mount Hebo Air Force Station kept losing battles with the weather, the nation's air defense and early warning system was in flux, too. A Russian bomber could cross the Pacific Ocean to deliver its lethal payload at 500 miles per hour in 1962. Nuclear warheads delivered by intercontinental ballistic missiles—the next-generation delivery system—can travel at speeds topping 15,000 miles per hour. The elaborate SAGE defense system gave way near the end of the Kennedy administration to a new the-best-defense-is-a-good-offense military doctrine, dubbed "mutual assured destruction" and known by the telling acronym, MAD.

The Mount Hebo Air Force Station mission began to change just before the third large radome was destroyed by wind. New radar was constructed to detect submarines capable of launching ballistic missiles. The 689th Radar Squadron was joined in July 1967 by an air force detachment from the Fourteenth Missile Warning Squadron. Searching the seas became as important as searching the skies, or more so. The nation's early warning system continued to evolve. When the Soviets in the 1970s developed submarines that could fire their nuclear arsenal from more than four thousand miles away, Mount Hebo and similar radar stations around the country grew obsolete. The Oregon base was deactivated in 1979–1980 and declared surplus for disposal in 1983. The site reverted back to the US Forest Service, but not before Oregon State prison supervisors visited to check out Mount Hebo as a possible prison site. Their conclusion? The environment was too harsh.

An 8.5-mile-long road winds through thick hemlock forests all the way from sea level to the top of Mount Hebo. Afternoon rain showers and sunbreaks battled for control of the weather on an early October day in 2013. The temperature dropped from 60 degrees Fahrenheit at sea level to 44 degrees on the mountaintop. Winds gusting to 30 miles per hour ushered in clouds that spit hail and rain and cloaked in soggy mist the five communications towers that stand guard at the end of the road. The droning of the tower's diesel-powered generators was muted by the whistling sound of the biting wind.

Two bearded, locally grown hunters drove by slowly with a four-point buck deer splayed out in the back of their battered pickup truck. We exchanged small talk about the panoramic view, not to be seen that day. Neither of these Oregon coastal inhabitants was born yet when the Columbus Day Storm made landfall. On the south side of a gravel road, a low-slung wooden rail fence bordered a grassy meadow knoll that spills off the top of the mountain to the south. An empty pint bottle of Jim Beam whiskey discarded by the fence suggested Mount Hebo hosts its fair share of revelry on starlit summer nights.

During those same short summers, the Mount Hebo meadows are alive with wildflowers, including broadleaf lupine, harsh paintbrush, and early blue violet. The coastal mountain is home to the last stable population of the endangered silverspot butterfly. On this October day, the meadows were neither colorful nor welcoming to visitors of any kind. Mount Hebo flashed a "no trespassing" sign, entombed in wind-wracked clouds that hid the panoramic views and masked stories from its Cold War past. The gnarly shore pines that dotted the meadow looked like beaten-down old-timers, clinging to life. I scurried back to the car, turned on the heater full bore, and started to drive away. A nondescript stone memorial by the side of the road caught my eye. The simple testimony to the men and women who lived and worked there read like military code: "In Memory of Those Who Served at Mount Hebo AFS, Oregon. 689th Radar Sq., October 1956–June 1979. Det. 2 14th MWS July 1967–September 1979." The downhill drive seemed much quicker than the drive to the top of the mountain. Isn't that the way when you retrace your steps? Back on Highway 101 heading south, the sky turned blue in between rain showers. It looked like the top of Mount Hebo could be peeking through the clouds. I did not turn around to confirm my suspicion. The winds swept off Mount Hebo and rushed twenty miles north into the lush, green Tillamook Valley, which is fed by five rivers, all flowing off the coastal mountains into the soggy valley before emptying into the bay and the Pacific Ocean.

Newlyweds Don and Andrea Jenck had just returned to their Tillamook dairy farm from a ten-day road trip to Southern California that included the obligatory tourist stops at Disneyland and Knott's Berry Farm. Their honeymoon memories were fresh in their minds, but about to be whisked away in the wind. They approached the outskirts of town from the south, crossing the Third Street Bridge over the Trask River, named after Elbridge Trask, the first white pioneer to venture to the shores of Tillamook Bay in 1848. They watched with a mix of fascination and fear as the wind struck a man walking along the road, blowing his raincoat over his head and sending him tumbling into a roadside ditch. The Jencks stopped to offer aid, but the windblown stranger dismissed them with an "I'm okay" and a wave of the hand.

"That's when I knew the honeymoon was over, I tell you," Jenck said with a heavy accent that still connects her to her native land of Belgium, which she and her family left behind when she was a young child. "That storm was the scariest thing I have ever seen in my life." It was about to get a lot scarier. Reaching home, Andrea scurried inside while Don, nicknamed Hooker, headed for the barn, where their seventy dairy cows waited to be milked. He sent half the herd to the milking parlor and went to work. The wind intensified and sent the barn doors banging. He thought the wind was going to tear the doors off the barn, so he left the milking parlor to close them. But he never had a chance. When he stepped outside the barn, the wind grabbed him and tossed him across the road and over a barbed-wire fence. He landed in a field softened by two weeks of heavy rain. Not injured, but stunned, Hooker picked himself off the ground. Before he could collect his thoughts, he heard a sickening, crashing sound as the barn collapsed, sending a billowing cloud of hay dust and splintered wood spiraling into the air. The collapsed barn knocked the power lines down along the roadway. The sound of bawling cows mixed with the moaning wind, a sound made more awful by the gathering gloom of darkness. About half the cows in the herd that remained to be milked were crushed to death or crippled when a hundred tons of hay that was stored in the hayloft above the stalls came crashing down. The Jencks had an old barn, and it was common in old barns to store the hay above the stalls. After the storm a lot of dairy farmers in the Tillamook Valley started storing their hay separate from their cows, Jenck said.

Hooker staggered back to the house. "Oh my God, the barn blew over," he told his wife. "The power is out and the wind was so loud. It was terrifying." The terror of collapsed barns and crushed cows was not isolated to the Jenck's farm. The Steiner dairy farm in Tillamook also suffered similar losses. More than half the herd of fifty-seven milk cows was crushed under a pile of hay and timber. Dairy farmers and their friends and neighbors were left with the gruesome task of climbing and crawling amid the tangled heaps of debris. Armed with rifles, they put their crippled cows out of their misery. "It was the most devastating thing I've ever seen," recalled Ron Zerker, who grew up on a Tillamook dairy farm not far from the Jenck place and witnessed some of the macabre cleanup and recovery the day after the storm passed. "Jenck was lucky he made it out of there alive."

The grisly cleanup work around the Tillamook dairies continued all weekend, recalled Christy Nicandri, a teenager living in the Oregon coastal community when the storm struck. She remembered stacks of dairy cows lying in the fields, waiting to be hauled away. Many of the farmers had put their cows out to pasture in advance of the windstorm. They were the lucky ones. Still, they faced the tricky task of milking their cows in the midst of power outages blamed on the storm. At first, Zerker, a teenager at the time, was relegated to the old-fashioned way of milking—by hand. Then he had a bright idea, one that would relieve his aching hands. Zerker screwed a garden hose onto a connector he threaded to a plug in the manifold of an old Chevrolet farm truck. He fastened the other end of the hose to the milking machine, which ran off air pressure. When he revved up the truck engine the milking machine kicked into action.

The livestock losses spread into the Willamette Valley, including in the Rickreall area near Salem. That's where Case Barendrecht was building up a herd of Holstein and Hereford cows and steers, but his plans were waylaid by the storm. About 4 p.m., 90 mile-per-hour wind gusts were buffeting his huge hay and stock barn, which housed about sixty cows and steers feeding on silage on the concrete main floor. Overhead, some three hundred tons of hay were stored for winter feeding. Suddenly, a barn wall collapsed and the building shifted ten feet to the north, sending the second floor, the hay, and the barn ceiling thundering down in a cloud of hay dust, timbers, and splintered wood. Many of the young cows died instantly, their heads and body parts protruding from the debris. Others were put out of their misery that night by Barendrecht and his rifle-toting neighbors, who also helped haul the salvageable hay to their barns and storage sheds, making room for a Polk County crane operator Sunday to move sections of the barn's second floor off the animals, piece by piece. Two live calves were found under the final section of flooring removed. One had a broken leg and had to be destroyed. Miraculously, one calf survived. The farm losses totaled some $100,000 at the time, or $775,000 in 2016 dollars. The farm was not insured.

The Jencks also lacked insurance, but they benefited from a stroke of good fortune the week after the storm. A fellow from Scappoose, Oregon, visiting friends in Tillamook, learned of the Jencks' plight and paid them a visit. He told them he was liquidating his son's dairy herd because of his son's sudden death during a trip to Switzerland. He had thirty pregnant heifers for sale, if the Jencks were interested. The Jencks said thanks for the offer, but told him they didn't have the money to buy the replacement cows. Don't worry about it, their benefactor said. Pay me back when the cows start producing milk. "So we took the cows, strictly on a handshake," Andrea Jenck recalled. "We were very thankful. The quality of those heifers was so good. Within two years we were the top-producing dairy farm in Tillamook County."

I visited the farm owned and operated by two of Andrea and Hooker's sons, Donald and Joseph Jenck, in 2013, on a typical early October day that offered rain showers, intermittent sun breaks, and gray-white cumulus clouds plastered against the black-green coastal mountains. Some two hundred Holstein cows lolled about and grazed in a rich green pasture not far from the site of the old barn that collapsed in 1962. Each cow carried a distinct pattern of black-and-white markings, not unlike the orca whales that cruise the ocean waters just west of the farm. What the mature cows have in common is an ability to produce six to eight gallons of milk daily for an average annual gross income of about $3,500 per year per cow.

These dairy facts are prominently displayed in town at the Tillamook County Creamery Association plant, which plays host to about one million visitors a year. But nowhere on the grounds is there any mention or display of the havoc done to Tillamook dairy farms by the devastating 1962 windstorm. There's not a word about the storm in *The Tillamook Way*, the 150-page account of the Tillamook creamery history written in 2000 by Archie Satterfield, celebrating North America's most successful farmer-owned dairy cooperative, which formed in 1909. The only 1962 event chronicled in the book is a rift between the co-op and the Tillamook Cheese and Dairy Association over Grade A milk. The battle, known as "the big split" lingered until 1968, when the creamery bought out the cheese and dairy association.

The storm didn't register in the creamery history book. Maybe the images of collapsed barns and cow carcasses littering fields don't belong there. But the storm remains etched in the mind of Andrea Jenck, who can gaze across a tranquil pastoral setting from her backyard to the scene of the collapsed barn and crushed cows. "Whenever we have a windstorm, it brings back memories, and I still get scared half to death," she said.

6
Ground Zero

If broadcast meteorologist Jack Capell and the other weather forecasters working in Portland late Friday afternoon could have seen the storm from space late Friday afternoon, they would have seen a monstrous, comma-shaped cloud with a tightly formed storm center advancing north at 40 miles per hour off the Oregon coast, spinning in a counterclockwise direction, covering a giant swath of the North Pacific Ocean. It would take a hurricane ten days to travel as far as the Columbus Day Storm did in thirty-six hours, a distance of some 1,800 miles.

Late afternoon Friday the strength of the tempest bore down on Corvallis, Oregon, home of Oregon State University, just an hour east of Newport in the central Willamette Valley, ground zero for the storm. The US Weather Bureau observer at the West Coast Airlines terminal at the Corvallis airport logged a wind gust of 127 miles per hour at 4 p.m. The wind was tearing the weather station and its instruments apart. The airport beacon toppled, and the anemometer was ripped to pieces. The next entry in his weather log, at 4:15 p.m., read "ABANDONED STATION." It was the only time a federal weather station in the Pacific Northwest has been abandoned because of stormy weather.

Damage to wind-reading gauges was commonplace along the path of the storm. It's safe to bet that the strongest winds in some locales occurred after the anemometers were destroyed. This suggests that the historical record of wind velocities is conservative. Exposed ridgetops and hills were no place to be during the Columbus Day Storm. Just ask Vern Johnson, who was eleven years old, living in a home with southern exposure on Logsdon Ridge, seven miles north of Corvallis, Oregon.

As the winds raked the ridge late that afternoon, Johnson was sitting on a sofa in the living room, scanning a Christmas catalogue that had just arrived in the mail. The windows were flexing in the wind and nails started

Toppled trees damaged the Van Buren Street Bridge in Corvallis, Oregon. Courtesy of the Oregon State University Valley Library Special Collections and Archive Research Center.

pulling loose from the open beams that supported the roof. He hollered to his mother and sister, telling them the house was coming apart. At first they didn't believe him. "I got up to leave and saw the roof start to lift off the beams," Johnson recalled fifty years later. "At that point I went outside and stood on the north side of the house. Next I heard my mother and sister screaming as the windows blew in, scattering glass across the living room."

His mother and sister fled the home and all three staggered to the west side of the house and hugged the ground. They witnessed the south half of the roof blow away and land about forty feet to the north. Huddled on the ground, they watched with awe as the walls, carport, and guest room collapsed. The home, just three years old, was demolished. After the winds subsided, Johnson remembers finding clothes and other household items tangled in the fence line about one-quarter mile north of their mangled home. With the help of thirty neighbors and friends, the Johnsons rebuilt their home in the weeks that followed, living with neighbors until the task was complete.

Another memorable storm scene played out twenty miles due north of Corvallis at the Oregon College of Education campus in Monmouth. Student Wes Luchau pressed a Graflex 4x5 press camera to his chest, waiting for the wind to die down enough that he could leave the entryway of the college administration building. It seemed like an eternity since the twenty-eight-year-old married student had snapped the photograph that would become the lasting image of the 1962 Columbus Day Storm.

Many dramatic photos were shot as the storm raged from Northern California in the morning until it finally petered out in southern British Columbia in the predawn hours of October 13. The photos depicted dairy cows crushed to death under collapsed, hay-laden barns from Tillamook to Rickreall, not far from the Monmouth college campus; cars flattened by falling oak trees at the Oregon State Hospital in Salem; and a massive Bonneville Power Administration transmission tower lying in a crumpled heap near the Washington-Oregon border. Each image tells a story of the wind's power and destruction, a mighty hand from the south pushing relentlessly north up the valley and into the Puget Sound lowlands and across the border into Canada.

However, most would agree Luchau's shot is the top attention-grabber of them all, freezing in time the exact moment when the steeple and bell tower on the south wing of Campbell Hall, which was built in 1889 at what is now Western Oregon University, separated from the oldest building in the state's public higher education system and toppled to the ground in a swirling cloud of brick and mortar dust. Other riveting storm photos capture the aftermath of destructive moments. A mix of good fortune, talent, and timing earned Luchau an iconic photograph of destruction as it happened.

When the storm struck campus in the late afternoon, Luchau, a married student with two young daughters, was studying in the life sciences laboratory on the second floor of the college administration building, across the street from Campbell Hall. Suddenly, the lights blinked out. Luchau peered out the laboratory window and watched in awe as a huge maple tree lifted up and fell over, its massive root ball emerging from the ground. Luchau ran downstairs and was met there by Wes Roberts, an air force enlistee and Monmouth resident who had stopped by the college campus for a look at the storm damage. He had been sent home from his construction job at the Adair Air Force Station near Corvallis in advance of the storm.

Just then, faculty member Don Mayo came running toward the two men from the Student Commons Building, where photography supplies were stored. Mayo knew Luchau was in a photography class and was assigned a student job responsible for all the photography and darkroom work for college publications. Mayo thrust the Graflex 4x5 press camera into Luchau's hands, along with two film holders and sheets of film representing four exposures. That's what Luchau had to work with as the wind roared and stately Douglas-fir trees in the college's storied Grove snapped and crashed to the ground. Luchau took a photo of the Grove, a tree stand planted in 1869 by the Christian founders of the college. Then he took another as the mix of more than fifty firs and maples started to disassemble, no match for the wind's fury. He had two shots left as he stood in the administration building entryway some fifty yards from Campbell Hall, a Gothic Revival–style building with a two-story tower. Roberts's gaze fixed on the bell tower on Campbell Hall. He noticed brick mortar starting to leak from the base of the twenty-foot-tall attachment to the hall's south wing. He alerted Luchau. A minute later a gust of wind lifted the tower and suspended it at a ten-degree tilt. Then the tower returned to its more than ninety-year-old footing. Luchau was feeling fortunate: he had caught the listing tower on film. He had one shot left.

Over the next two minutes the tower rocked in a gust of wind, then rocked again and leaned to a 15-degree angle. Luchau edged out to the sidewalk, but resisted the urge to take a picture, knowing he had only one exposure left. Roberts followed close behind, and grabbed Luchau around the waist to steady him in the wind. Then came the moment every photographer dreams of, that right place–right time experience that can't be planned. Watching through his viewfinder, Luchau saw the tower start to topple. He waited a split second for the tower to reach a point of no return.

He pushed the shutter for the final shot, just before the tower crashed to the ground, releasing a cloud of debris that billowed into the wind-whipped sky. There was no clanging of a bell. The bell was long gone, removed in 1920, along with a flagpole, to deter student pranks. He stood staring at a heap of rubble that moments before had been the emblematic Campbell Hall bell tower. In his hand was the photo of a lifetime. This is the stuff photographers' dreams are made of, he thought to himself.

Luchau started to walk to the Student Commons Building to find Mayo, but the winds were too strong, nearly knocking him off his feet. The last thing he wanted to do was fall and damage the camera and its precious contents. He turned back to the administration building and hunkered down with Roberts, hoping the relentless winds would subside. After what seemed like an eternity, maybe ninety minutes or so, the winds did begin to relent. Luchau set off on foot to share his treasured photo with Mayo. When Mayo learned what Luchau had captured on film, he grabbed the camera, jumped in his old Studebaker sedan and, in the eerie calm after the storm, headed for Salem and the office of the *Oregon Statesman*, a morning newspaper and one of the two dailies in the capital city. The twenty-mile trip was a hazardous one. The streets were littered with downed trees, arcing power lines, shattered glass, and building debris. Mayo was determined to deliver the goods. He knew what a stunning photo he had in his possession.

He made it safely to the newsroom, but it was steeped in darkness, victim of a widespread power outage. Undaunted, the staff there figured out a way to develop the film by the light of a Coleman lantern. The picture appeared on the front page of the paper the next day. "I thought that was kind of neat and that that was the end of it," Luchau recalled.

Guess again. Associated Press, the world's oldest and largest news-gathering agency, transmitted the picture across the country and around the world through its wire photo service. Within twenty-four hours, the image had appeared in dozens of news outlets in the United States and abroad. Luchau sensed his photo had gone viral when he got a call from his brother in Germany. He had seen the photo, too.

Four days after the storm, Luchau received a letter from Oregon College of Education president Leonard W. Rice, an Idaho farm boy and coal miner's son who was valedictorian of his 1941 graduating class at Brigham Young University in Utah. He praised Luchau for his photography skills in capturing the fall of the historic Campbell Hall bell tower. "Although I have been at OCE less than three months, it has been long enough to appreciate the symbolic position the tower played in the history of Oregon College of Education and sense the loss that students, graduates and faculty feel," Rice wrote. "You captured a once-in-a-lifetime shot. Congratulations. The picture did full honors to the death of the tower."

A few days later, AP sent Luchau a check for $25. Soon thereafter, Luchau was contacted by officials from *Life* magazine. A meeting with magazine officials in Portland, Oregon, ensued, and Luchau sold the picture for $400. Luchau saved a copy of the check, emblazoned with the name Time, Inc., which purchased *Life* magazine in 1936. The photo appeared in *Life* magazine two weeks after the storm, accompanied by these introductory words: "Like an invisible demon howling out of the depths of hell, the winds clawed at the campus hall." Anyone with twenty cents to spare could own the magazine and enjoy Luchau's shot of a lifetime.

Life magazine was still in its heyday in the early 1960s, read by millions each week and filled with larger-than-life photos and stories of President John F. Kennedy and his family, movie starlet Elizabeth Taylor, the fledgling Apollo space program and the burgeoning Vietnam War. There in the fall of 1962, Luchau for one issue joined the ranks of such world-renowned *Life* photographers as Alfred Eisenstaedt, who captured the 1945 V-J Day picture of a nurse and sailor locked in a kissing embrace in New York City's Times Square. Eisenstaedt's photographic masterpieces graced the cover of *Life* magazine eighty-six times. Luchau's picture didn't make the cover; it was trumped by an illustration for a new magazine series on the human body. But his Columbus Day Storm picture swept across two inside pages, good enough to land Luchau a sizeable check. What did Luchau do with the money? He bought a twin lens reflex camera and a high fidelity stereo. "Here we were, living in a tiny little apartment on campus, and we had some of the best stereo and camera equipment in town," he said.

Luchau shared his memories of the storm and the photo during a two-hour interview in the kitchen of the old farmhouse he remodeled on the eastern, rural, side of Salem's urban growth boundary, the side that grows ornamental grasses, beans, and nursery crops instead of strip malls and apartments. A tall man with disheveled hair, bushy eyebrows, and a jutting jaw, Luchau was happy to pull out a black briefcase filled with Columbus Day Storm photo memorabilia. Included inside was his last copy of the October 26, 1962, *Life* magazine. He used to have a dozen copies, but he's given away all but one over the years.

There are newspaper clippings from Birmingham, Alabama, to New York City with front-page photos of the falling tower. His photo anchored the front-page storm coverage of the state's largest newspaper, *The Oregonian* in Portland, the day after the storm. There was the letter from college president Rice and another from Wendell Webb, managing editor of the *Oregon Statesman* when the storm struck, thanking and congratulating Luchau for his outstanding photo. "It was a real smash," Webb said in his October 18, 1962, letter. "It was mighty fine of you and Mr. Mayo to get it over to us, too. Normally, we reimburse to the extent of $5 per picture. But this certainly is a special case. I am glad to enclose one for a bit more."

Luchau accepted the $25 check with modesty and gratitude, the notoriety of the moment he captured on film just beginning to sink in.

Luchau, a kind and easygoing man, still savors the memories of that fateful day. I asked him what it's like to be forever known as the guy who took the iconic Columbus Day Storm photo. He shrugged and called it his fifteen minutes of fame. He was also quick to credit Mayo and Roberts for their helping hands. "I was a lucky guy," he said. "I just happened to be there, looking at Campbell Hall when Don [Mayo] brought me the camera." Equipped with a new camera, Luchau remained a devoted amateur photographer after the storm. Friends and relatives hired him to photograph weddings and graduations, and Luchau toyed with the idea of becoming a professional photographer. But he opted instead to complete college and earn his teaching certificate, which led to a seventeen-year career as a middle-school science teacher in Salem, followed by many years running a trucking business and helping operate a rural Salem blueberry farm owned by his wife's family.

The Oregon College of Education in Monmouth was marred one other way by the Columbus Day Storm winds. On the eventful fall Friday night, the college football stadium was set to host a high school football game between the hometown Central High School and Cascade High School of nearby Turner, Oregon. Ray Coleman was athletic director at Cascade and one of his duties was to make sure the field was ready for play on game day. He headed for the college campus around 4 p.m. He stopped at a Mobil gas station on Main Street and talked to the owner, "Shorty" Fisher, the mayor of Monmouth. Fisher told Coleman about radio reports he'd just heard, warning of winds up to 100 miles per hour shooting up the Willamette Valley from the south and lasting right into game time. Coleman absorbed the grim weather report, thought "better safe than sorry," and called officials at both schools to cancel the game.

Good thing he did. The stadium press box, which could hold up to forty people, blew away in the storm. "Had it been occupied, I along with other coaches, press reporters, film crew, statistician, scouts, etc., could have been seriously injured," Coleman said. Elsewhere in the Willamette Valley, school officials with less information than Coleman had allowed football games to be played. Near Eugene, Oregon, Monroe and Coburg High Schools squared off in a homecoming football game at 3 p.m., just as the storm was reaching the lower Willamette Valley. The goal posts at one end of the field blew down early in the contest, but the game went on. Just before halftime, the winds tore apart the small wooden grandstand, sending the homecoming queen and her court scrambling for safety. Still, the teams continued to play. Early in the third quarter, a fifty-gallon drum flew across the playing field, just missing the players. The drum jumped a fence and landed in a neighboring farmer's field. Finally, the game was suspended, with Monroe leading twenty to nothing.

A contest between Mohawk and Detroit high schools near Salem, Oregon, had its fair share of unexpected adventure. During halftime, spectators watched in awe as the wind flattened the nearby bus barn at the school. Detroit scored its only points of the game when a bad snap flew over the Mohawk punter's head and sailed into the end zone. Final score: Mohawk, twelve, Detroit, two.

Punting footballs was an adventure at any Friday night football game attempted in the teeth of the storm. In the Seattle area, the storm didn't arrive until games were well under way. Shoreline High School punter Tom White lofted a kick into the wind from his twenty-seven yard line. The football reached the thirty-five, climbed skyward, then reversed direction and flew back to Shoreline's twenty yard line. White caught his own punt and was tackled by an Edmonds player on the seventeen yard line for a ten-yard loss. Later in the first half—the game was postponed at halftime— White kicked a wind-aided punt sixty-one yards.

The southerly winds shooting up the Willamette Valley found an easy target on a small hilltop east of Salem, Oregon, some sixteen miles northeast of Monmouth. Perched on the hill was Oregon State Hospital, an imposing Victorian behemoth built in 1883 and named the Kirkbride Building after Thomas K. Kirkbride, the 1800s Quaker physician whose model for "insane asylums" of the day was a soothing country setting, filled with physical and mental activities for the patients. The hospital setting was anything but soothing at the peak of the storm. Even today, the old hospital and its 148-acre parklike campus feel the full force of whatever wind may blow, from the light one that ruffled my clothes and licked at Old Glory flying flexed and fluttered from a flagpole outside the hospital on a late May day in 2013 to the deadly blast of hurricane-force winds that visited the hospital and all its vulnerable residents in the early evening hours October 12, 1962.

The hospital has a storied place in Pacific Northwest history. When no other mental hospital on the West Coast would, it welcomed actors Michael Douglas, Jack Nicholson, and the rest of the Academy Award–winning crew and cast from *One Flew Over the Cuckoo's Nest* to film on the hospital grounds for eleven weeks in 1975. The movie is based on the best-selling novel by Ken Kesey, an Oregonian, counterculture icon, and literary giant of the 1960s who used his LSD-induced experiences during 1959 research at a VA hospital in Palo Alto, California, to describe life in a mental hospital through the eyes of a person with schizophrenia.

Kesey's novel, which came to movie life at the Oregon State Hospital, was published in 1962, the year of the Columbus Day Storm. The commanding views from the hospital grounds are a sight to behold: look to the west and you can see down Center Street almost all the way to downtown Salem and the winding, north-flowing Willamette River. "They'll put you out at the end of Center Street if you don't watch out," Salem people said in the 1930s, half in jest, half serious when someone started acting strange in town. "They'll put you out in the bughouse with the rest of the nuts."

To the east, it's only a few miles to the urban-rural divide, that sharp land-use boundary drawn by legislators who supported Oregon's landmark growth management act pushed by Governor Tom McCall and signed into law in 1973. The national leader in trying to curb suburban sprawl, Oregon worked hard to keep urban-density housing and strip malls from spilling into the farmlands and open space that extends all the way to the foothills of the Cascade Range to the east and the Coast Range to the west.

The northern view from the old hospital grounds crosses Center Street and is quickly blocked by mostly abandoned, depressed-looking hospital buildings that reflect a decades-old image of 3,545 patients crowded into dozens of hospital wards, sleeping two in rooms designed for one, cots and beds spilling out into the hallways. To the south stands the Oregon State Penitentiary. Over the years, prisoners became hospital patients and hospital patients became prisoners in an uneven attempt to punish misdeeds at the hospital and treat prison inmates ruled criminally insane.

Presiding over the hospital and its some 2,500 patients in the fall of 1962 was Dr. Dean K. Brooks, one of the most notable and well-liked superintendents in the hospital's long and checkered history. He served at the hospital helm from 1955 to 1982. A Kansas native and graduate of the University of Kansas Medical School in 1942, Brooks was hired as a

psychiatrist at the hospital in 1947, the year the first lobotomy was per-
formed there, a practice halted the year he became superintendent. He was
a lifelong advocate for his patients, embracing them as unique individuals
with their own hopes and dreams, albeit sometimes clouded or distorted
by mental illness. "Find fact, not fault," was the mantra that governed
Brooks's professional and personal life.

He let patients shed their hospital-issued clothes in favor of their own
street attire. An avid rock climber—he used to rappel down the hospital
exterior—he took patients and hospital staff on two-week-long summer
excursions into Oregon wilderness areas. He didn't hesitate to allow certain
patients he screened to babysit and play with his three young daughters,
who grew up on the hospital grounds in the two-story superintendent's
house, which was built in 1909. Informal by nature, he insisted patients
and colleagues call him Dean. His patients-first creed showed in many
ways. For instance, he established a patients' committee to meet with him
weekly to air grievances and discuss hospital policies. The only rules he
enforced at the hourly sessions were no smoking and no petty gripes.

Brooks is most remembered for welcoming the crew of *One Flew Over
the Cuckoo's Nest* to film the acclaimed movie at the hospital. Brooks took
the invitation one bold step further, playing the role of Dr. Jack Spivey,
the mental hospital director whose ad-libbed hospital admission dialogue
with rebellious patient Randle P. McMurphy, played by Jack Nicholson, is
one of the film's highlights.

I interviewed Dr. Brooks on May 24, 2013, at the Willson House, a
Salem nursing home not far from the Oregon State Hospital. At ninety-
six, and despite failing eyesight and the need for an oxygen tank, he
remained an advocate for the mentally ill, arguing for better treatment in
community-based settings. He voiced frustration over the swelling ranks
of the mentally ill who are homeless or jailed. He'd suffered a nasty fall five
days before my visit, and was bruised and weakened by the accident. But
he still greeted me warmly, and struggled from his bed into a chair. He
corrected me when I called him Dr. Brooks—"It's Dean." Then he reached
back in time to October 12, 1962.

Brooks was driving back to the hospital from a meeting in Salem at
the Fairview Training Center for the developmentally disabled late that
afternoon when the wind picked up. His apprehension grew when he
saw a large sign fly through the air past his car. He made it home all right

and remembered his wife reminding him that the windstorm would most likely knock out power at the hospital. He started thinking about what that would mean to the patients.

As dusk descended on the hospital grounds, the mighty winds ripped through the hospital campus. The winds blew down more than ninety-two oak, maple, and fir trees, many of them several feet in diameter and eighty to one hundred feet tall. One of the falling oaks crushed a patient, Timothy O'Sullivan, thirty-five, as he returned to his ward from the hospital cafeteria, where the patients had been fed a semi-cold meal in the gathering gloom. O'Sullivan was rushed to the University of Oregon Medical School Hospital in Portland, where he died from his injuries one week later. Other falling trees flattened cars belonging to hospital employees.

Another patient, Fredrick Johnson, forty-eight, was one of the many who had ground privileges, which meant a patient was free to leave the hospital for trips to town. "Johnson had been downtown and hurried back to the hospital through the storm," Brooks stated in a November 1962 storm report to state officials. "He got to the ward, laid down on his bed and died immediately of a heart attack. There had been no previous history of a coronary problem."

The scene Johnson had witnessed downtown was enough to scare anyone to death. Newspaper accounts of the frightful evening describe downtown pedestrians and shoppers blown off their feet by the winds as they, too, tried to scurry home to safety. Some of those who stayed upright were hit by glass from shattered storefront windows. Cars were blown onto sidewalks and yards, and the steeple at St. Paul's Episcopal Church toppled and speared the ground. Walls of downtown buildings, including the Marion Motor Hotel and the Capital Press building, collapsed from the force of the winds, and the roof blew off the nursery at the Salem Memorial Hospital. All the beds and babies were moved to another floor, recalled Salem resident Judy Brown, who worked at the hospital the night of the storm.

Floyd McKay, a young reporter at the Salem morning newspaper, the *Oregon Statesman*, recalled scrambling around town past downed power lines and fallen trees to gather storm stories and photographs of the damage, returning to a newsroom that operated by candlelight and flashlight. Covering the story was hazardous duty, said McKay, whose long career in journalism included stints as a news analyst at KGW television in Portland—in a newsroom that still included Capell—and chair of the journalism department at Western Washington State University in Bellingham, Washington. On his way back into town from his house, McKay saw an old, vacant mill building that looked like it would be blown away in the storm. Thinking that would make a good picture, he grabbed his camera and stationed himself upwind of the mill, next to a railroad car sitting on a railroad spur line that served the mill.

"After a bit, I heard a roar and saw a sheet of metal siding shoot under the rail car at a high rate of speed, literally a few feet from where I was standing, protected by the wheels of the rail car," he said. "If I had been a few feet away, the sheet would have cut off my legs. I waited for a lull in the wind and got the hell out of there. The mill never did blow apart; it was so rickety the wind blew right through it." The Oregon state capitol campus took a weather beating, with dozens of downed trees, colored windows smashed in the state supreme court building, and the 3.5-ton bronze *Circuit Rider* statue knocked upside down off its pedestal by a falling spruce tree. Willson Park, a dense grove of old-growth Douglas-fir trees near the state capitol building, donated to the city by Salem pioneer Dr. William Willson, was all but destroyed by the storm. It took a crew a month, working seven days a week, to clean up the damage. The park was replanted and made more pedestrian friendly, but for many it was never the same.

First-term governor Mark Hatfield, forty, was in the midst of a reelection campaign when the storm struck. The popular Republican governor, with the steadfast antiwar beliefs born of his Baptist faith and World War II experiences, was visiting INPLY, a plywood plant in Independence, Oregon, just a few miles southwest of the state capital. He went on the mid-afternoon plant tour with just plant officials and his security guard, leaving the Polk County Republican dignitaries stewing outside. Not long after that, the governor suddenly exited the building, indicating he had just received word of a severe windstorm approaching rapidly from the south. He returned to his office in the capitol building, joined by officials from the National Guard, Oregon State Police, and civil defense. The governor and his ad hoc emergency management crew pieced together a picture of the storm damage through calls to local officials, radio reports from state agencies, and from Capell and the news team at the KGW radio station in Portland. The governor declared a state of emergency in Western Oregon at 5:50 p.m., Friday, and put 655 National Guard troops on alert, most of

A falling spruce tree knocked the 3.5-ton bronze Circuit Rider statue off its pedestal on the state capitol campus in Salem, Oregon. Photo by Verdi and Thelma Walser.

them dispatched to guard against looting in Portland, Salem, Albany, Cottage Grove, Newport, and Newberg.

Back at Oregon State Hospital, the winds continued to pick apart the J Building, one of the original structures on that small exposed hilltop. Limbs and broken-off treetops sailed through the air, smashed windows, punctured the historic building's cupola, and bounced off the stiff brick exterior. Nearly a full acre of metal roofing was ripped away by the winds. A 1901 horse barn that had been converted into an equipment-storage building was destroyed in the storm. The hospital suffered storm damage far greater than any other state institution in Salem—more than $500,000 by today's measure.

At the peak of the storm, which raged through Salem in the early evening hours with winds topping 90 miles per hour, Brooks's thoughts turned to the patients and their needs. "I went ward to ward, asking them what they wanted the most," he recalled, his eyes locked on a long-ago memory. "They said they wanted light." So Brooks bent the hospital rules to fit the demands of the emergency: he allowed candles on the wards, to bring some light to the darkened nooks and crannies of the rundown hospital.

Brooks paused a moment as he harkened back to that hellish night when flickering candlelight cast eerie shadows in the dozens of hospital wards as the vicious winds howled and screamed outside. "I'll tell you this—the place was huge," Brooks said as his shoulders sagged with the memory of all that responsibility he carried with him for decades. "But I'm proud of how the patients responded—they were wonderful." There were, however, a few exceptions. Three patients in a medium-security ward in the J Building jumped from second-story windows in apparent attempts to escape in the wake of the Friday night storm. One of the men suffered leg, foot, and back fractures, another fractured an ankle, and a third sprained an ankle. The men, all in their twenties, were quickly controlled by hospital attendants.

As the winds began to die down, the rain picked up, creating a new set of challenges. "It was necessary to evacuate six wards on the top floor of our main building," Brooks wrote in his storm report. Some thirty patients

The morning after the storm, Oregon State Hospital superintendent Dr. Dean Brooks was on the hospital roof, helping with repairs. Courtesy of the Brooks family.

in a maximum-security ward on the third floor could not be moved to other areas of the hospital for safety reasons. Someone suggested transporting them to the neighboring state penitentiary. That idea was nixed in favor of doling out raincoats and slickers to the patients to wait it out until emergency roof repairs could begin the next day.

Power was restored to the hospital around midnight. And throughout the ordeal, the hospital stayed heated, thanks to an emergency generator. The next day broke clear and mild, with a mid-October sun shining down on a main hospital building stripped of its roof. Hospital employees reporting to work Saturday had to navigate a ghastly tangle of downed trees and limbs. One hospital employee, Jeannette Simpson, tripped on woody debris and fell, breaking her right hip and forearm. And Dean Brooks showed up on the J Building rooftop the next day to help with temporary roof repairs.

Not long after the storm, Brooks read Kesey's debut novel. "What was your impression of the book?" I asked him. "I hated everything about it," Brooks said. But his attitude changed over the years as he stopped viewing the book as an indictment of mental health hospitals and more as an allegorical tale of institutional abuse of power. "It could be a bank, a school, the army, or a hospital where the story takes place," he came to say.

In the early 1970s, Brooks polled patients and staff alike about the prospects of inviting the movie crew to film at the hospital. He even had patients query movie producer Michael Douglas about his intentions. "The patients asked, 'If we're in the movie, will you make us look like nuts?'" Brooks recalled. "Douglas said 'no.'" Then the patients asked, "Will we get paid? Douglas said 'yes.'"

In a 1973 excerpt from his diary, on display at the hospital museum, which opened October 2012 in the old Kirkbride Building, Brooks mused on the pros and cons of the decision: "Of course there will be criticism, misunderstanding, uptight people, etc., but they'll happen anyway—and I think Oregon would gain something by being gutsy enough to cooperate in telling a very good story, done with genius." The project was approved, and filming began in January 1975. Some ninety patients and hospital staff had paid roles in the movie.

The movie generated money for the hospital in other ways, noted Ray Tipton, former hospital director of support services. The movie producers paid the hospital $4,000 a day to use their on-site wood, machine, and metal shops to build props for the movie, everything from the broom Chief Bromden pushes around the halls of his ward to the hydrotherapy unit the chief throws out the window to escape after he suffocates the once-defiant, but then lobotomized, McMurphy. Both props are on display in the museum. "That money went directly into a fund to benefit patients," Tipton said. Hospital staff who owned cars vintage 1955 to 1962—the movie is set in 1963—were paid $25 a day anytime their vehicles were used in movie scenes. "I had a 1962 green Dodge pickup that appears in the movie," Tipton said. "I made a nice chunk of change."

Douglas and Nicholson both returned to the hospital to visit with patients after the movie racked up five Academy Awards for best picture, best director, best actor, best actress, and best screenplay. Douglas is a major supporter of the museum. "It was an important memory in my life," he's quoted as saying in the museum exhibit. "It totally changed my life both personally and my understanding a little bit about mental health." In 2015 he called it the most profound movie he ever helped shape.

Louise Fletcher, who plays the diabolical Nurse Ratched in the film, shares with Brooks the same birthdate—July 22. She returned to the hospital for the museum's grand opening, which coincided with the fifty-year anniversary of the mighty windstorm, and spent some time reminiscing with Brooks. Brooks, who appears in the movie in several sequences, even joined the Screen Actors Guild. He also basked in the glory of the film's success. A director's chair Douglas and coproducer Saul Zaentz gifted to Brooks is part of the museum exhibit. "Brooks was a little bit vain, but a wonderful guy," Tipton recalled. "He was a hundred years ahead of his time—a giant in the mental health community."

My interview with Dean Brooks lasted about forty-five minutes. It tired him out, but I'd like to believe in a good way. He grabbed my hand and smiled as we said our goodbyes. Then I followed his youngest daughter, Dr. Ulista Brooks, who took me on a driving tour of the Oregon State Hospital where she grew up as a child, volunteered as a teenage candy striper, and still works as an internist. Most of the buildings were empty and neglected, replaced by a new hospital that had 579 patients and 1,769 employees in January 2013. She showed me her family's old home on the campus grounds, and pointed to the bedroom window where she watched a giant maple tree fall over in the 1962 windstorm's initial assault. At first she thought someone was cutting the tree down, which infuriated her. Then she realized it was the work of the storm, a sudden source of chaos and destruction at a hospital ground that has been her childhood home, playground, and workplace for more than sixty years.

7
A Wind Like No Other

The Columbus Day Storm's march through the Pacific Northwest earned it a lofty standing in the annals of nontropical windstorms to strike in the lower forty-eight states. Ted Buehner, a weather warning meteorologist based at the National Weather Service office in Seattle, minced no words when a newspaper reporter in October 2012 asked him to rank the 1962 storm. It was the "strongest non-tropical windstorm to ever strike the lower 48 states," he said. Buehner said the experience as a young boy in Portland living through the storm aroused his interest in severe weather and guided him to a career devoted to weather.

Pacific Northwest weather guru Cliff Mass, a professor who has taught and researched weather science in the University of Washington's Atmospheric Sciences Department since 1981, paid his respects to the storm at an October 11, 2012, gathering of weather buffs at the university's Kane Hall. "By all accounts, the Columbus Day Storm was the most damaging windstorm to strike the Pacific Northwest since the arrival of European settlers," said Mass, who places the storm in another special niche in his 2008 book, *The Weather of the Pacific Northwest.*

Mass loves to dissect and talk about weather and, much like his mentor Carl Sagan, share his knowledge with the public. He is next to reverential when he talks about the Columbus Day Storm, and enjoys pointing out how much stronger it was than the October 31, 1991, tempest off the coast of Massachusetts known through book and movie fame as "the Perfect Storm." For instance, the Columbus Day Storm notched winds ranging from 60 to 170 miles per hour. The Perfect Storm weighed in at 60 to 109 miles per hour. The Perfect Storm hit Maine. The Columbus Day Storm swept through Northern California, Oregon, Washington, and southern British Columbia. Columbus Day Storm damage weighted to modern times totaled at least $1.86 billion, with some estimates climbing higher than $3 billion when timber damage is included. The Perfect Storm losses reached around $346 million.

A day after the storm's fifty-year anniversary, members of the Oregon Chapter of the American Meteorological Society gathered at a public meeting in Portland to pay homage to the epic storm. The slide show presented by Tyree Wilde, warning coordination meteorologist for the National Weather Service in Portland, was titled, "Columbus Day Storm: The Benchmark for Pacific Northwest Windstorms." His slide show included this description of the storm shortly after it had passed through the region: "The most powerful 'non-tropical" windstorm to ever hit the lower 48 states in recorded American history struck the Pacific Coast."

Christopher Burt, a meteorologist and national historian of severe weather events, couldn't let the Columbus Day Storm's fiftieth anniversary pass without offering this weather disaster tribute: "This storm ranked as one of the most deadly and destructive in that region's and United States history." National Weather Service weather forecasters in Washington State ranked the top ten weather events of the twentieth century in the Evergreen State, and placed the Columbus Day Storm at the top of the list, calling it "the mother of all wind storms this century, the wind storm all others are compared to and the strongest widespread non-hurricane wind storm to strike the continental United States this century."

Then Superstorm Sandy arrived on the scene, changing the calculus and conversation around historic storms. When Sandy made landfall on the New Jersey shoreline October 29, 2012, just north of Atlantic City, it had shed its identity as a hurricane after a week of lethal, destructive behavior in the Caribbean Sea. Sandy's new East Coast persona was eerily similar to that of the Columbus Day Storm on the other side of the continent fifty years prior.

The Columbus Day Storm of 1962 and Superstorm Sandy of 2012 both grew from tropical cyclones—called typhoons in the Western Hemisphere and hurricanes in the Eastern Hemisphere. They both transitioned into midlatitude cyclones, switching energy sources from warm tropical waters to severe differences in temperatures and barometric pressures between colliding warm and cold weather systems. The storms switched from hot-blooded to cold-blooded when they both mixed their decaying tropical energies with blasts of cold air from the Arctic. They were both extratropical storms, which means the center of the storms were colder than the surrounding air. The storms had fronts—boundaries between air masses of different densities. And extratropical storms are much larger than hurricanes—from seven hundred miles to more than a thousand miles across, compared with two hundred miles to five hundred miles for a hurricane.

Both storms made landfall with hurricane-force winds, but the Columbus Day Storm winds were stronger than Sandy's. Winds topping 100 miles per hour were logged for at least nine locations along the path of the Columbus Day Storm. In many places power outages and wind-damaged gauges may have left the strongest winds undetected. By comparison, the peak gust for Superstorm Sandy was 96 miles per hour, reported at Eaton Neck, Long Island. But Superstorm Sandy topped the Columbus Day Storm on several other measures, including a low barometric pressure reading of 28.01 Monday night at the Atlantic City Airport, compared with a low landfall reading of 28.50 in Newport and Corvallis late Friday afternoon of the Columbus Day Storm.

In addition, Superstorm Sandy had destructive storm surges, including a nine-foot storm surge and 13.88-foot tide level at the southern tip of Manhattan. The Columbus Day Storm lacked storm surges of significance, in part because of its northerly path, almost parallel to the coastline. Sandy's wrath was felt in twenty-four states, compared with the three states and one Canadian province visited by the 1962 big blow. Property damage and fatalities were higher in 2012, too, as one would expect of a storm striking the nation's most populated area, compared with the lightly populated far corner of the country in 1962. Propel the Columbus Day Storm forward fifty years and it's possible, with all the population growth and development in the Pacific Northwest, that the death toll and monetary damage could start to rival that of Superstorm Sandy's mark in the United States—seventy-two direct deaths and $65 billion in damages.

Even after Superstorm Sandy left its mark, Mass called the winds associated with the Columbus Day Storm perhaps the mightiest to make landfall in the lower forty-eight states, and the total storm profile the strongest of any to strike the West Coast in recorded history. The Columbus Day Storm of 1962 was a freak of nature, a weather outlier, a beastly wind that caught weather forecasters flat-footed and dumbfounded. Armed with sophisticated weather-predicting tools and computer models, Superstorm Sandy and all its menace was forecasted days in advance. The Columbus Day Storm hit like a sucker punch. It was poorly understood and underestimated. It was sudden and unexpected. In 1962, weather reports on television were just leaving the era of cartoonish graphics, with wind personified as long-haired old men, exhaling tempests. Weather forecasting tools were crude—not much more than reports from ships at sea. The coastal radar weather stations in place recorded only rainfall, not wind. The weather satellites orbiting the earth were in their infancy, offering little, if any, visual information of typhoons and hurricanes forming in the oceans in 1962. There was no Doppler radar, a tool to detect wind speed and motion, along the nation's coast to warn of advancing windstorms. Weather buoys at sea, floating weather stations that measure air temperature, wind speed, wind direction, and barometric pressure, were rare.

Even as wind damage mounted in the early afternoon in Southern Oregon, the US Weather Bureau stations in Portland and Seattle did not update their morning forecasts, leaving it to Portland broadcast meteorologist Jack Capell to sound the first accurate, albeit belated, alarm.

Wolf Read, a Seattle native and climatologist based in Vancouver, British Columbia, has researched major Pacific Northwest windstorms, and written about them, in painstaking detail. Like so many weather enthusiasts, it was a particular storm that first grabbed his attention, turning a curiosity about weather into a lifelong obsession. He was ten years old when a February 13, 1979, windstorm destroyed a 3,200-foot-long section of a bridge spanning Hood Canal, an extension of Puget Sound tucked against the east slope of the Olympic Mountains in Western Washington. Wind gusts on this fjord-like, westernmost arm of Puget Sound reached 100 miles per hour. "Before that, I was interested in hurricanes and tornadoes, but I had no idea such strong windstorms hit the Pacific Northwest," Read said years later. As a teenager living in Seattle, he spent countless hours at the NOAA National Weather Service office on the shores of Lake Washington, poring over old weather reports. Then as a graduate student at Oregon State University in Corvallis, Oregon, in 2002, he started digging into old newspaper accounts and records kept by the US Army Signal Service, a predecessor to the US Weather Bureau, and found a whopper of a storm that hit the Pacific Northwest on January 9, 1880.

That wind and snow storm, in the sparsely populated Washington and Oregon Territories, claimed three lives; knocked down barns, sheds, and houses from the Oregon coast to Seattle; and toppled half the trees in what later became Clark County in southwest Washington, with winds gusting to 90 miles per hour. The winds were succeeded by severe snowfall measuring three feet and more in the Puget Sound region. Thanks to Read's research, a late-nineteenth-century storm finally got its due. He dubbed it "Storm King." If cyclones competed for "strongest storm on record," in the Cascadia region, the final round would probably be between the windstorm of January 9, 1880, and the Columbus Day Storm of 1962, Read has often said.

So which storm would win the heavyweight title? "I am inclined to put the Columbus Day Storm on top," he said. Here's why: The Columbus Day Storm had much greater reach. Moving in a south-to-north trajectory, it raced through Northern California, Western Oregon, Western Washington, and southern British Columbia, covering some seventy-five thousand square miles before running out of steam. Storm King made landfall somewhere south of Astoria, Oregon, then beat a path northeast across northern Oregon and southwest Washington before breaking up in the Cascade Range. Storm King featured wind gusts that probably topped 100 miles per hour on the Oregon coast and 70 miles per hour in Portland. It set record-low barometric pressure readings at Astoria, Oregon (28.45) and Portland, Oregon (28.56). But the Columbus Day Storm logged 100-mile-per-hour-plus winds at multiple locations, from Cape Blanco in Southern Oregon to Renton, just south of Seattle. It destroyed or damaged more than fifty-three thousand homes. In the Willamette Valley damaged homes, barns, and outbuildings were the rule, not the exception.

The Columbus Day Storm behaved more like a hurricane than a mid-latitude cyclone, Read said. Retired National Weather Service meteorologist George Miller—he rode out the storm in the weather bureau's Portland office—agreed: "The large number of trees over a thousand years old that blew down in the storm suggests it wasn't just the storm of the century, but the storm of several centuries." Read's windstorm curiosity led to another analysis of the Columbus Day Storm he did while working on his doctorate in forest science climatology at the University of British Columbia in Vancouver, Canada. Presented at the October 13, 2012, American Meteorological Society meeting in Portland, Read's analysis compared the 1962 storm to other significant Pacific Northwest cyclones to make landfall from 1940 through 2000. He found four big hitters striking in 1945, 1962, 1981, and 1995, and five lesser storms in that period. His research suggests that a storm with winds of about 52 miles per hour and gusts exceeding 70 miles per hour can be expected every ten years in the region. A windstorm with average speeds of about 58 miles per hour can be expected about every forty years.

The 1981 windstorm was actually back-to-back weather systems, a storm characteristic shared with the Columbus Day Storm. That November 13–14 event featured winds gusting the first day near 100 miles per hour on the Oregon coast and 70 miles per hour in Western Washington and Oregon. The storms claimed thirteen lives, including four when a US Coast Guard helicopter crashed near Coos Bay while searching for a fishing boat.

Mass assessed modern-era Pacific Northwest windstorms dating back to 1960. In *Weather of the Pacific Northwest*, he ranked a January 20, 1993, storm as the third-most damaging in the past fifty years. On the day Arkansas native William Jefferson Clinton was sworn in as the forty-second president in "the other Washington," the Evergreen State was slammed with winds that topped 100 miles per hour in exposed areas of the Cascades and gusts of 80 miles per hour or more on the Washington coast and western interior. Six people died, seventy-nine homes and four apartment buildings were destroyed, and some 870,000 electric customers were cast into darkness.

Number two on the Mass power storm index since 1960 goes to the Hanukkah Eve storm of December 14–15, 2006. The winds gusted from 70 to 90 miles per hour through much of Western Washington, topped 100 miles per hour in the Cascades, and lashed the Oregon coast with hurricane-force winds, too. This storm approached on a more westerly track than most, claiming thirteen lives in Washington State and triggering a power outage that struck 1.3 million customers in Western Washington.

Of the ten attention-grabbing storms of the past seventy-five years analyzed by Mass and Read, only one produced widespread gusts of 80 to 110 miles per hour in the population centers of Western Oregon and Washington, and that was the Columbus Day Storm. The magnitude of the Columbus Day Storm gusts was about two times the wind force of the other major storm events in recent Pacific Northwest history, Read concluded. Looked at another way, the lowest peak gust in the Willamette Valley during the 1962 storm was 86 miles per hour at Eugene. That's higher than any maximum wind measured in the Willamette Valley from twenty-one windstorms from 1950 to 2003.

Then there are the storms from another time, a time when Pacific Northwest Indian tribes lived without interference from white man. For Pacific Northwest tribes, especially those ensconced along the coast, where storms born over the Pacific Ocean first make landfall, the legendary creature Thunderbird seems to embody almost every kind of severe weather—thunder and wind from his beating wings and lightning flashing from his eyes. Oral histories from the Hoh and Quileute tribes of the Olympic Peninsula speak of epic battles between Thunderbird and killer whales. When they fought, they would uproot trees and strip the timber off the mountainsides, which is the same result one would expect from hurricane-force windstorms.

The written accounts of explorers and pioneers to America's last frontier speak of severe weather events, including windstorms. Evidence of a major windstorm impressed Captain John Meares, a British fur trader and explorer who sailed through the Strait of Juan de Fuca in the summer of 1788, rounded Cape Flattery, and named the highest peak in the Olympic Mountains Mount Olympus. After peering at a windblown landscape amid otherwise dense conifer forests clinging to the coastline and stretching into the river valleys, foothills, and mountain slopes beyond, Meares observed,

> The force of Southerly storms was evident to every eye: large and extensive woods being laid flat by their power, the branches forming one long line to the North West, intermingled with roots of innumerable trees, which had been torn from their beds and helped to mark the furious course of their tempests.

The white explorers and pioneers kept coming, and so did the windstorms. Journals kept by members of the Lewis and Clark Expedition speak of a major windstorm striking their encampment southeast of Chinook Point on the north side of the mouth of the Columbia River on November 22, 1805: "The wind increased to a Storm from the S.S.E. and blew with violence.... O! how horrible is the day," William Clark wrote in his journal. He said the wind threw river water over the camp, adding to heavy rain that fell all day. Patrick Gass's journal entry had this to say: "The wind blew very hard from the south, and river was rougher than it has been since we came here. The wind and rain continued all day violent." Although similar in track to the Columbus Day Storm of 1962, the 1805 storm was unlikely to have been as fierce, weather historian Wolf Read said. It's a conclusion born in part by the fact there is no mention from the 1805 storm observations of falling trees.

The Hudson's Bay Company, a London-based fur-trading enterprise, built the first European settlement on the shores of South Puget Sound in 1833, just north of the Nisqually River, which flows off a glacier of the region's towering, omnipresent volcano—Mount Rainier. It wasn't long before this eclectic population of a dozen British, Hawaiian, French Canadian, West Indian, Native American, and American settlers learned firsthand that the temperate climate of the Pacific Northwest is punctuated by the occasional severe windstorm.

William Kitson, the Hudson's Bay factor, or head merchant, at the fort wrote in his journal about an October gale that tore apart the twelve-foot-high stockade just erected at the fort, and upended haystacks, too. In the middle of a severe winter of snow and sub-zero temperatures, the wind returned on February 2, 1834. "Toward day break this morning we were visited by a dreadful hurricane of wind which tore up some of the largest trees by the roots, broke others and nearly blew down the fort which was only saved by the shelter of the woods to windwards and the props we placed to support it," Kitson wrote.

The deadliest, and perhaps most freakish, windstorm on record in the Pacific Northwest was a sudden, southwest gale off the mouth of the Columbia River near Astoria on May 4, 1880, just shy of four months after Storm King struck the same area. The springtime windstorm caught hundreds of commercial salmon fishermen off guard, trying to row their wooden, twenty-four-foot boats back over the treacherous Columbia River bar to safety. But the river flow of the mighty Columbia, boosted by an early snowmelt, overwhelmed the flood tide the fishing fleet typically relied on to help them home with their catch. Instead of being pushed back into the river and safety by the flooding tide, the small boats were swept into massive breakers that formed over the Columbia bar. The death toll was not well documented, but estimated at anywhere from several dozen to 350.

The west side of the Olympic Peninsula was blindsided January 29, 1921, by a windstorm of epic proportion, one that notched wind gusts of some 150 miles per hour on the north side of the mouth of the Columbia River, then swept north with fury, leveling 6.8 billion board feet of ancient

Douglas-fir, western redcedar, hemlock, and spruce trees—20 to 40 percent of the timber on the northwest side of the Olympic Peninsula. US
Forest Service officials at that time said it was the most timber ever blown
down in a single windstorm in the nation's history. Yet it's less than half of
the downed timber credited to the Columbus Day Storm.

Another major windstorm struck the Pacific Northwest on October
31, 1934. This storm generated twenty-foot waves in Puget Sound and
the Strait of Juan de Fuca. It claimed twenty-two lives in Washington
and Oregon, including five fishermen lost in the turbulent seas near Port
Townsend. In Seattle, the winds tore the Pacific liner *President Madison*
from its moorings, causing it to smash and sink two other ships. Seattle
firefighters logged the busiest day in the history of the fire department,
dousing wind-whipped fires throughout the city.

The scientific and anecdotal weather record for the Pacific Northwest
is clear: Pacific storms bearing hurricane-force winds strike every few
decades, killing people, destroying personal property, causing extended
power outages, and blowing down the very evergreen trees that define the
region. Each storm has left its own unique mark—a record wind or an
all-time-low barometric pressure reading at a given locale. But no storm
in its totality comes close to the overall scope, wrath, and damage of the
Columbus Day Storm. The same holds true with a review of the 106 major
windstorms to strike Southern California from 1858 to November 2013.
There have been occasional gusts of wind topping 100 miles per hour in
isolated areas. On September 24 and 25, 1939, a tropical storm just downgraded from a hurricane before it made landfall had sustained winds of 50
miles per hour in the San Pedro, California, area and at least forty-eight
people died from sinking boats. Still, the Columbus Day Storm ranks at the
top of all West Coast extratropical windstorms.

Several factors came into play to elevate that storm above others in the
weather annals of the Pacific Northwest. On the morning of October 12,
1962, the warm moist air remnants of Typhoon Freda mixed with cool
air from the Gulf of Alaska and warm, moist tropical air from the south
as it raced eastward along a jet stream aimed at Northern California. The
storm was undergoing rapid, explosive cyclogenesis, also known as a meteorological bomb. To qualify as a bomb, a storm's central pressure must
drop rapidly, for instance, at least one millibar per hour over the course

of twenty-four hours (1,013 millibars is considered standard sea level pressure). In the case of the Columbus Day Storm, pressure at the storm center two hundred miles west of Eureka, California, at 9 a.m., October 12, had plummeted more than 22 millibars in just three hours as the storm pushed north and slightly east. The low-pressure reading at the storm's center—960 millibars and possibly a bit lower—is equivalent to what to expect with a category 3 hurricane and goes a long way toward explaining the record high winds.

The storm plunged into a trough of low pressure near the West Coast and was flung around the base of the trough on a distinctly northern path up the coastline. The storm center was close to land—within fifty miles of Astoria, Oregon, at the peak of the storm, which is as much as 150 to 200 miles closer to the coast than most cyclone storm centers that reach the Pacific Northwest. This is important because the strongest winds in a storm such as this are typically nearest to the storm's center and are positioned on the right side of a cyclone spinning counterclockwise in the Northern Hemisphere. This placed the Columbus Day Storm's ferocious winds over the region's most populated areas in the Willamette Valley and Puget Sound.

The west–east orientation of the pressure gradients, caused by the southerly flow of winds, also helped keep the winds funneled between the Coast Range and the Cascade Range, creating a Venturi effect of higher wind velocity moving through a constricted area. "If the winds had come from the west, the pressure gradient would have changed and the damage would not have been nearly as severe," said Kathy Dello, deputy director of the Oregon Climate Research Institute, a network of some 150 climate scientists with its main office at Oregon State University in Corvallis, Oregon. "We have extratropical storms visit us frequently. But the intensity of the low pressure, combined with the direction of the storm, and our topography, made this one historic."

Adding to the storm's might was another atmospheric condition: the upper atmosphere and surface-level winds were both blowing from the south. The lack of a stable layer of air between the strong winds aloft and the surface winds allowed the turbulent eddies from the winds aloft to mix with the surface winds, especially along the storm front. A more colorful explanation of the unstable air aloft was offered by a 1963 report in the *Western Conservation Journal*, a forestry trade magazine: "Unstable layers overhead boiled like a witch's cauldron; injecting a shot of energy equal to many an H-bomb to kick the storm in the pants as it chewed up the countryside."

The surface winds were enhanced by the speed of the storm, which traveled at 40 to 50 miles per hour. And the sheer size of the storm led to winds that sustained over areas for two to four hours. Combine that with unusually gusty winds and the storm damage mounted. The Columbus Day Storm arrived early in the storm season, which means the deciduous trees of the Pacific Northwest were heavy with leaves and more prone to wind damage. Parts of the Pacific Northwest had already received their normal allotment of rainfall for the month of October when the storm struck. The soils were saturated, making trees more susceptible to blowdown. In addition, cyclones such as the Columbus Day Storm go through birth, growth, maturity, decline, and death. When the big blow made landfall and moved across the Oregon coast and Willamette Valley, it was in the powerful prime of its life.

Fifty years after the Columbus Day Storm ripped through the Pacific Northwest, meteorologists, including Steve Pierce, president of the Oregon Chapter of the American Meteorological Society still marveled at the storm's might: "There has yet to be another tempest that even comes close to the furor of the Columbus Day Storm," Pierce remarked. "To this day the storm can best be summarized with words such as 'frightening' and 'amazing.'"

Climate scientists have studied and the public has pondered: Will climate change in the decades ahead increase the Pacific Northwest likelihood of more nontropical windstorms as strong as, or stronger than, the Columbus Day Storm? The vast majority of climate scientists agree that the earth is warming under the influence of greenhouse gas emissions, much of it generated by the burning of fossil fuels. Without a leveling-off of greenhouse gas emissions, especially carbon dioxide, by mid-century, Oregon could heat up, on average, by 3 to 7 degrees Fahrenheit by the 2050s and 5 to 14 degrees Fahrenheit by the 2080s, according to a January 2017 climate assessment report by the Oregon Climate Change Research Institute.

Increased flooding, sea-level rise, reduced snowpack, and diminished summer stream flows are all feared consequences of a warmer Pacific Northwest, the report stated. But the future for extratropical windstorms akin to the Columbus Day Storm is far less clear. The report noted an upward trend in the intensity and frequency of extreme storm events since 1950 in the Northern Hemisphere, but not in the Pacific Northwest. The research cited by the institute, led by Russell S. Vose, chief of the climate science division at NOAA's National Climatic Data Center in Asheville, North Carolina, offered this uncertain view of the future: "Neither climate model projections nor our understanding of the physical climate system leads to any conclusive answers regarding extratropical cyclone activity in a warming climate."

Climate scientists Christian Seiler and F. W. Zwiers, who conduct research for the Pacific Climate Impacts Consortium at the University of Victoria, British Columbia, prepared a paper that appeared in 2015, in the publication *Climate Dynamics*, titled, "How Will Climate Change Affect Explosive Cyclones in the Extratropics of the Northern Hemisphere?" They identified and tracked cyclones from twenty-three state-of-the-art climate models for the recent past (1981–1999) and modeled the future (2081–2099).

"Explosive cyclones are projected to shift northwards about 2.2 degrees latitude on average in the northern Pacific, with fewer and weaker events south of 45 degrees N [which runs through northern Oregon] and more frequent and stronger events north of this latitude. . . . These changes also cancel each other out when averaging across the entire basin," the scientists go on to say. The northern shift in cyclone activity relates to a projected poleward shift in the jet stream, the main transport vehicle for Pacific Northwest windstorms. One potential dampening effect on nontropical windstorm intensity in the Pacific Northwest noted by Seiler and other climate scientists is something called polar amplification, which refers to the Arctic region losing sea ice from global warming and heating up at a faster rate than the midlatitude regions, weakening the extreme differences in temperatures that fuel cyclones in the North Pacific Ocean.

Presenting at the Northwest Climate Conference at the University of Washington in September 2014, UW atmospheric scientist Cliff Mass offered his scientific assessment to the question of whether climate change will amplify the number and intensity of damaging Northwest windstorms in the future. "There's no reason to expect Northwest windstorms to be more frequent, or more intense," Mass said, relying on an analysis of a suite of global computer models collectively known by the eye-glazing name of "coupled model intercomparison project 5 (CMIP5)", the same models Seiler and Zwiers used to produce their paper. Mass also referenced in his August 31, 2015, weather blog the findings of the 2012 International Panel on Climate Change, which reported low confidence in the ability to link global warming to changes in the regional intensity of midlatitude cyclones.

At the same time, Mass, Seiler, and others have noted there is a wide degree of variability in the climate models, which speaks directly to the difficulty of predicting the future of extreme Pacific Northwest windstorms. "There's no evidence that extratropical cyclones in the Pacific Northwest will intensify significantly with climate change, but a storm such as the Columbus Day Storm could surely happen again," said the Oregon climate institute's Dello.

8
Fallen Forests

The Columbus Day Storm swept through some of the mightiest conifer forests anywhere on earth, part of a continuous sylvan that once stretched uninterrupted from the fjords of southeast Alaska to the fog-enshrouded redwood forests of coastal Northern California. In their natural state, the softwood forests of the Pacific Northwest were, and are, unsurpassed anywhere in the world for density and tree size. Three of the five tallest tree species in the world are found along the path of the storm: Douglas-fir, Sitka spruce, and redwood. Together with two other evergreen cone-bearing trees—the western hemlock and western redcedar—and deciduous trees such as alder and bigleaf maple, the forested landscape confronted by the storm helped define the storm's might.

When a major windstorm rips through the far northwest region of the United States, it breaks trees in half, snaps them off at the base or uproots them from the ground. Send those same winds through the plains of Nebraska or badlands of the Dakotas and it would just be wind rushing forward, unimpeded and less destructive. The ubiquitous, towering trees of the Pacific Northwest become the force multipliers, the measure of the might of any windstorm. In the urban and suburban areas along the path of the Columbus Day Storm, the trees smashed cars, and buildings, blocked roads, left city parks and neighborhoods denuded, and caused multiple fatalities. In the forests of the Pacific Northwest, the fallen trees posed an historic, transformative challenge to the timber industry.

Forestlands represent roughly 50 percent of the land mass in Washington and Oregon. About half of the timbered area in the two states felt the wrath of the storm. The volume of tree loss from the storm was unprecedented. An estimated fifteen billion board feet of timber fell victim to the winds that swept through the Pacific Northwest that day and night. A board foot is a measure of the usable wood in a tree, equivalent to a piece of lumber one foot long, one foot wide, and one inch thick. The Columbus Day Storm windfall contained enough lumber to frame nearly one million homes. It's a volume more than twice what loggers extracted from the public and private forests of Washington and Oregon in 2014. It was more than three times the timber destroyed by the explosion of Mount St. Helens in May of 1980, and a greater loss than what occurred during the two most destructive wildfires to strike the region in modern times—the Yacolt Burn in southwest Washington and Oregon in September 1902 and the Tillamook Burn that began in August 1933 near the northern Oregon coast. The tree toll for those two big burns was twelve billion board feet each.

By the time the Columbus Day Storm confronted the region, the age and density of the signature tree species of the Pacific Northwest had changed dramatically. The centuries-old Douglas-fir forests that grew on flat ground and gently sloping hillsides below 2,000 feet elevation had already been logged and replaced by smaller trees. The Sitka spruce stands found along the Oregon coast had been exploited to meet the World War I appetite for straight-grained spruce used to build biplanes for the war effort. An old-growth inventory of old trees in Washington State, defined as trees more than two hundred years old, covered 40 percent of the state's forested area twenty years before the epic windstorm. Some forty-five years after the storm, ancient forests were down to less than 12 percent of the forestland, the vast majority of it in national parks and wilderness areas.

The landscape was fractured and discordant in 1962, composed of a creeping checkerboard pattern of clear-cuts and haphazardly managed plantations of younger trees owned by private timber companies and state and federal agencies, all united under the common goal of trying to meet the nation's insatiable appetite for wood. The forestlands available for logging butted up against isolated tracts of thick, dark ancient forests that soaked up the rain and blotted out the sun, some in wilderness and park designation, and some still caught in the tug-of-war between the timber industry and those who wanted to preserve the remaining pristine forests in their unaltered state.

Douglas-fir (*Pseudotsuga menziesii*) was in 1962 and is today the dominant conifer in the forest, representing about one-third of the total forestland acreage in Washington and Oregon. It is a heavy-limbed evergreen, reaching diameters of ten feet or more, towering to heights exceeding two hundred feet and living for five centuries and beyond. Named after early-nineteenth-century British explorer and botanist David Douglas, this tree is the mainstay of the timber industry and serves as the state tree of Oregon. It features deeply furrowed bark, deep blue-green foliage, and spirals of needles that hang like pendants from the branches. Douglas-fir has a paradoxical identity despite its abundance and beauty: it has never had a universally accepted common name. Early botanists called it Oregon pine, but eventually it was formally classified as a false hemlock and named in honor of its earlier discoverer, Scottish physician Dr. Archibald Menzies, who made note of the tree on Vancouver Island in 1793 as a member of British captain George Vancouver's exploring expedition. Coast Salish tribes burned the green boughs of the Douglas-fir in their sweat lodges and used the heavy, durable wood to build fires. Douglas-fir timber is milled into many forms for trusses, beams, dimension lumber, flooring, and posts. Douglas-fir is the preferred wood for manufacturing plywood.

Sitka spruce (*Picea sitchensis*) doesn't take a back seat to any of the grand conifers, growing quicker than others, reaching heights above 250 feet and ages approaching a thousand years. Growing close to the marine shores of the Pacific Northwest, the strong yet lightweight wood was logged extensively along the Central Oregon coast by the biplane builders of World War I. Musical instrument makers worldwide covet the wood.

Western hemlock (*Tsuga heterophylla*) is a slow-growing, smaller conifer, a member of the pine family and the Washington state tree. It features smallish, abundant cones; shorter, more delicate needles; and droopy treetops. The western hemlock can dominate a forest devoid of the ravages of wind and fire. Native Americans found several uses for the bark. They cooked it to treat tuberculosis, rheumatic fever, and hemorrhaging. They ate the boiled inner bark as a survival food in the winter. They used dark dyes from the bark to coat their fishing nets and fishing lines, making them less visible to fish. Once disdained as a weed tree by the timber industry, it has grown in value and stature over the past one hundred years.

Western redcedar (*Thuja plicata*) is a member of the cypress family and revered in the cultures of the indigenous people of the Pacific Northwest. Ancient western redcedar can measure more than fifteen feet in diameter and grow to more than two hundred feet tall. It's the tree of dugout canoes and totem poles, shelter and clothing, baskets and boxes. It stands out in a forest of conifers, with its stringy cinnamon-red to grayish-brown bark, bulging trunk, aromatic fernlike boughs, and broken multi-leadered tops. The wood is durable and versatile, used for home construction from the shakes on the roof to the planks on the deck.

The Columbus Day Storm generated a Herculean response to salvage the wood. A log export market to Japan and other Asian countries grew up overnight. The network of logging roads in the region's forests grew by thousands of miles just to reach all the downed timber. Timberland owners embarked on a new style of timber management to plant and grow trees faster to buffer the timber industry against the random chaos and destruction wrought by wind and fire.

Unsuspecting loggers, timber surveyors, heavy equipment operators, and others employed on the public and private forestlands of the Pacific Northwest on Friday October 12 saw the winds pick up, sending small limbs flying through the air, then knocking larger limbs and entire trees to the ground. Many of them, including US Forest Service employee Carl Anderson, a timber sales administrator on the Shelton Ranger District in the Olympic National Forest, had to act quickly and decisively to outwit the winds.

Anderson was driving his small pickup early Friday evening near Vance Creek in the southeast corner of the Olympic National Forest when the winds started knocking down Douglas-fir trees across the road in front of him and behind him. He drove uphill on a spur road to a thirty-acre clear-cut and parked in the middle of it to avoid the falling trees. He watched waves of trees blow over along the edge of the clear-cut. Concerned his truck might blow off the mountainside with him in it, he threw logs and boulders into the truck bed and cab to add ballast. Later that night he finally remembered to radio his coworkers at the Shelton Ranger District in Hoodsport, Washington, to report his location and situation. They'd been trying to reach him, and were worried for his safety. After an anxious night in the truck, he started walking down the logging road. He had to climb over, under, and around hundreds of old-growth Douglas-fir before he was picked up by a Simpson Timber Company employee.

It didn't take long Friday night for Glen Hawley, management forester for the Washington State Department of Natural Resources' Kelso District, to realize the windstorm striking the twin timber towns of Kelso and Longview would be a game changer for him. First, he watched the wind

Many glass storefront windows like this one in Longview, Washington, were shattered by the Columbus Day Storm winds. Courtesy of the Cowlitz County Historical Museum, Kelso, Washington.

blow out the living room windows of his neighbor's house in Kelso—the single-pane windows found in homes and stores in 1962 were no match for the hurricane-force winds.

Then Hawley's thoughts turned to the state forestlands he managed in southwest Washington, a forestry district bordered by US Forest Service and Weyerhaeuser Company land, home to the largest stands of old-growth state timber left in Western Washington, much of it east of Interstate 5 in the river valleys flowing off the slopes of the conical-shaped Mount St. Helens. He sensed that the severe windstorm had wreaked havoc on an unprecedented scale. When he woke up Saturday morning from a fitful sleep, he called his boss and said, "I don't suppose I'm going to be going hunting this year." Hawley guessed right. He would be preoccupied with timber salvage operations created by the storm for the next three years.

Public and private timberland owners wasted no time responding to the natural disaster. The Northwest Forest Pest Action Council's Timber Disaster Committee, which represented a who's-who of the timber industry, convened a meeting in Portland, Oregon, three days after the storm struck. Their goal: survey wind damage on some twenty-five million acres of timberland by January 1963 and formulate a plan to get the downed timber out of the woods and into the marketplace as soon as possible.

In addition, President Kennedy convened a timber summit in Portland October 30–31 to help forge an aggressive federal logging plan, which was then unfettered by forestry rules and regulations that grew in complexity and controversy in the 1980s and beyond. The tenure of the summit was a far cry from the forest summit President Bill Clinton would preside over in Portland on April 2, 1993, when his administration searched for a path forward to end a gridlock in the region's national forests. By then it had become a conflict that pitted wood-hungry loggers demanding continued access to the national timber supply against environmentalists trying to protect the dwindling old-growth forests and the creatures that depend on those ancient stands of trees, including the northern spotted owl and marbled murrelet. In 1962, there were no environmental protections in place to govern forestry. Timber harvests were directed by simple guidelines that set annual harvest goals and required landowners to flag their timber sales and respect property boundaries. "It was before the national and state environmental policy acts were adopted," recalled Kenhelm Russell, a DNR forester who chaired the timber disaster committee meetings in 1965. "The good old boys went to the back room, had a couple beers, and decided what to do."

There was a sense of urgency in the air with all that wood on the ground. Timberland owners were in a race with the Douglas-fir beetle, a one-eighth-inch-long, six-legged, oblong creature that, if left to its own devices, would bore into the fallen and broken timber in the spring of 1963 to lay eggs in tunnels in the cambium layer, the place of tree growth located between the bark and wood, rich in sugar and nutrients. When the eggs hatched, the beetle larvae would eat their way out of the tunnels at right angles and emerge as young adult beetles in the spring of 1964, ready to devour standing timber, too. Fresh in the minds of many foresters was a 1951 windstorm that struck Western Oregon hard and triggered a beetle epidemic that claimed three billion board feet of timber. Luckily, when the Columbus Day Storm struck, the Douglas-fir beetle population was at one of its lowest levels in years. The other threat—catastrophic wildfires fueled by the fallen timber—also weighed heavy on the timber industry.

On the rain-soaked Siuslaw National Forest, a 650,000-acre forest that stretches along the Oregon coast and into the Oregon Coast Range south to north from Coos Bay to Tillamook, storm damage survey work began early Saturday morning, the day after the epic windstorm. Neil Phillips, a US Forest Service employee on the Siuslaw's Waldport Ranger District, remembered walking into the woods that day because the roads were still blocked by fallen trees. He and a coworker trekked ten miles Saturday and nine miles Sunday, climbing over, under, and around blowdown, mapping and estimating the damage. "The powers that be approved a dawn-to-dusk, seven-days-a-week work schedule that communicated the seriousness of the situation," said Waldport district timber sales planter Ken Roberts. "We responded with enthusiasm."

The first timber sale on the Siuslaw was advertised on October 16, just four days after the storm. The blowdown on the national forest was estimated at 740 million board feet, more than any other national forest in the Pacific Northwest and a volume that dwarfed the forest's annual allowable harvest of 334 million board feet. Damage was most severe on high ridges exposed to the wind. A swath of the Siuslaw some twenty miles wide, one

Tangled masses of broken and downed trees dotted the Siuslaw National Forest after the Columbus Day Storm, including this example in what is now the Oregon Dunes National Recreational Area. Courtesy of the Oregon State University Valley Library Special Collections and Research Center, Corvallis, Oregon.

hundred miles long and about fifteen miles inland from the coast suffered the brunt of the damage, Phillips said. It wasn't just trees that fell on the national forest. The fire lookout living quarters at Marys Peak, which is the highest point in the Coast Range at 4,097 feet, blew away in the storm.

Timber blown down or shattered by the storm in DNR's Kelso District in southwest Washington was estimated at 880 million board feet, which was a whopping twenty-two times more than the annual harvest rate. "It was heady times, exciting times," Hawley recalled. "There was a lot of pressure to prepare those timber sales, but I had a determined, focused work force. It was the most important occurrence of my professional life."

Foresters went to work preparing timber sales, which required them to calculate the fallen timber and flag boundaries for timber sales and harvests, a job called timber cruising. It was hazardous work in a jumbled, chaotic landscape. "A lot of times you were ten or twenty feet off the ground, walking on fallen trees and limbs," recalled DNR forester Roy Friis, who worked much of the time in the shadow of Mount St. Helens. "You could cruise all day without your feet touching the ground. In the old-growth stands there wasn't too much brush, but the limbs from that blowdown—holy mackerel!"

The 1950s had seen an explosion in logging road construction in the forests of the Pacific Northwest as timber harvests moved past logged-over shorelines and lowland river valleys and moved onto higher-elevation ground. Logging trucks had replaced the historic progression of horses, oxen, steam donkeys, railroad spur lines, and log-rafts floated down rivers as the main way to get fallen trees from the forests to the mills. The Columbus Day Storm touched off another round of road-building. In Weyerhaeuser's St. Helens Tree Farm in southwest Washington, the logging road inventory grew from 270 miles in 1950 to 1,300 miles one year after the storm. About 70 percent of the three billion board feet of storm-damaged timber on Weyerhaeuser property was not accessible by roads that existed when the storm struck.

Every tree farm, national forest, and tract of state timber needed new roads to get the wood out. On the Siuslaw National Forest, a 1,200-mile network of logging roads prior to the storm grew by 450 miles to reach the fallen timber. Many of those hastily built logging roads would come back to haunt the environment, and the industry. Logging roads boost the likelihood of landslides, ten to one hundred times over. The roads

Mount St. Helens in all its pre-eruption, symmetrical beauty looms over a storm-damaged forest landscape in southwest Washington. Courtesy of Roy Friis.

sliced out of the hillsides became conduits during rainstorms in the ensu-ing years, delivering tons of sediment to rivers and streams, smothering salmon spawning habitat and choking stream channels. By the late 1990s, the miles of logging roads in Oregon's national forests had reached more than seventy-three thousand, triple the footprint that existed in 1960. The 1990s were also a time when private, state, and federal timberland owners began spending tens of millions of dollars to retire orphan and substan-dard roads, many built just before and after the Columbus Day Storm.

Loggers faced hazardous conditions in the blowdown areas. Trees ripped from the ground by the winds featured root wads held together with rocks and dirt. Cutting into the base of the tree, where some of the highest-quality wood grows, could turn root wads into lethal flying debris. Or it could snap the stump upright with enough force to maim or kill a logger. The wind-ravaged forests were filled with splintered trees and huge broken limbs, called "widow makers," lodged high above the ground in standing trees, waiting to crash on unsuspecting loggers who had to pick and choose carefully what to cut. About 25 percent of the volume of trees damaged or knocked down in the storm stayed in the woods, unfit for har-vest. "Fortunately nobody working for me was seriously hurt," Hawley said.

Finding a market for the surplus wood that emerged from the ravaged forests was a huge challenge. It was more wood than the mills of the Pacific Northwest could handle. The Pacific Northwest timber industry cast its collective gaze across the Pacific Ocean to Japan in search of a market hungry for wood. The storm struck just as Japan's appetite for wood was growing. Japan's post–World War II economy was on the rise, expanding at the rate of 10 percent per year. Japanese forests were depleted, but Japan had invested in sawmill capacity in the years following the war. Japan needed logs, not finished wood products, to fuel those mills and produce lumber for construction of Japanese homes traditionally built from wood. The Columbus Day Storm and the glut of timber it knocked down made it Japan's perfect storm.

The Weyerhaeuser Company, a dominant force in the region's timber industry since 1900, wasted no time moving raw logs across the Pacific Ocean to Japan, especially hemlock, which faced a weak market in North America but a welcoming one in Japan. "In export, hemlock was the preferred species and demanded a higher price," noted George Weyerhaeuser, great-grandson of Frederick Weyerhaeuser, the founder of the giant timber company. "So the log trade was a godsend." Log exports had just begun to flow from the Pacific Northwest to Japan when the storm made landfall, reaching 375 million board feet in 1961. The storm was a potent shot in the arm for the log export market, one that grew steadily through the 1960s, reaching two billion board feet in 1968. By 1988, Pacific Northwest log exports to Asia from state and private forestland topped 3.5 billion board feet per year. The birth of the log export market is one of the storm's most enduring legacies, although a controversial one at best. Small mill owners, environmentalists, and others decried the loss of raw material to feed local mills. Longshoremen, timber companies, teamsters, and free trade advocates countered that the wood drew a higher price in Asia than it did back home.

Malcolm Dick, chief forester of Weyerhaeuser's Twin Harbors Tree Farm in the Willapa Hills and Grays Harbor region of storm-battered southwest Washington, was busy after the storm, mapping out emergency timber harvests and traveling to Japan to set up markets for the glut of fallen timber. But he found time to take his teenage son, Bob Dick, out to a ridgetop overlooking the North River, in what had been a forest-carpeted landscape in the rolling Willapa Hills between the Olympic Mountains to the north and the Columbia River to the south. "I remember it like it was yesterday—there was nothing standing for miles," Malcolm's son said more than fifty years after the storm. "Mother Nature can change things overnight. That deeply impressed me." The awestruck teen went on to a career in forestry, serving as Alaska's chief forester before a long stint as an Olympia, Washington–based lobbyist and consultant for commercial timber interests.

The storm also served as a rude awakening for the timber industry. Trees that remained from before the white man's axe was wielded, and trees that had grown up to replace the fallen ones, were flattened by the wind. The Columbus Day Storm was an equal opportunity destroyer. The storm created isolated and splotchy, dramatic and continuous, voids in the forest, enough to shake landowner faith in long-range harvest plans and timber inventories. "The Columbus Day Storm jump-started the industry," Dick remarked. "It forced landowners to really start investing in their land."

Out of the storm grew the concept of high-yield forestry, the idea that timberland owners could grow more trees, and faster, if they beefed up their seedling production using genetically superior seed stock grown in controlled, seed-orchard settings. The new forestry template also called on the timber industry to fertilize the forest soils, thin the young tree stands to promote vigorous growth, and use herbicide sprays to combat brush and alder trees in the forest understory. Weyerhaeuser paved the way, and other timberland owners followed. By 1975, Weyerhaeuser's high-yield forests in the Pacific Northwest were growing double the wood found on unmanaged forests.

High-yield forestry had its detractors. A September 1974 article in *Audubon* magazine, by John G. Mitchell, a well-respected conservationist and writer, summed up questions critics had of the new way of growing trees in a Weyerhaeuser profile with the telling title, "The Best of the SOBs": "They want to know about Weyerhaeuser's long-term commitment to the land, about soil erosion and stream sedimentation and the widespread application of herbicides and nitrogen fertilizers and what risks have been considered in the development of genetically selected seed stock," Mitchell surmised. "They want to know why, when US home-builders raise alarms of a timber famine and timbermen cry for increased harvests in national forests, Weyerhaeuser must export saw-logs by the million board feet to Japan."

The fallen forests of the Columbus Day Storm were more than just a messy, time-consuming cleanup chore for the timber industry. All the downed and damaged trees did more than kick-start a log export debate that lasted for decades or fuel a rash of road-building in the woods or give birth to more aggressive management of private and public timberlands. From a forest ecology point of view, high-powered windstorms and runaway forest fires are part of the natural order in the woods. These natural disasters create biological diversity across the landscape. Trees broken by the wind become snags that house cavity-nesting birds. Timber left to rot on the ground provides habitat for thousands of plant and animal species. The decaying wood also rebuilds the forest soils for the next generation of trees. The open spaces created by windthrow allow sunlight into the forest, promoting new vegetative growth and adding a mix of tree ages and tree species to the landscape.

There are many ways to contemplate the forest disturbances caused by wind. Think of the old-growth trees in the forest as gigantic community platforms. For instance, high in the canopy of ancient Douglas-fir trees, decaying needles create thick mats of soil that support thousands of species of plants, vertebrates, insects, and fungi. When the tree falls, those communities collapse. Another fascinating element of forest life plays out underground in what's become known as the Wood Wide Web, the vast network of interconnected tree roots and fungi that serves as the pathway for transporting water, nutrients, and carbohydrates among the trees. Recent research points to a symbiotic relationship between mature, or mother, trees and the seedlings that they feed through the Wood Wide Web. When the mother tree falls victim to wind, fire, or the logger's chain saw, the survival rate of the nearby seedlings can be reduced. Just as the Columbus Day Storm wreaked havoc on human neighborhoods and communities, it had a similar effect on the forest communities as well.

9

The Wind and Wine

A few minutes before 5 p.m., October 12, 1962, the clocks stopped ticking in the Dundee Hills thirty miles southwest of Portland, Oregon. Home to southeast-facing filbert, walnut, and prune orchards, the Dundee Hills confronted the violent winds of the Columbus Day Storm head-on. The storm's arrival in the heart of nut and fruit orchard country would help usher in a new agricultural era in the Willamette Valley, unbeknownst to the shell-shocked farmers who saw their orchards decimated in two or three hours on an October harvest-season day.

The fruit and nut trees, planted in rolling countryside that stretched past the small farm communities of Layfette, Dayton, Dundee, and Newberg in Yamhill County, were no match for winds that gusted across the exposed hillsides at 100 miles per hour. Those trees bent, twisted, and fell, many uprooted, others clinging to the moderately rich, red, well-drained soils deposited fifteen million years ago by rivers of lava flowing down the Columbia River Gorge and emptying into the Willamette Valley from erupting volcanoes on the east side of the Cascade Range.

A tally of the storm damage by Yamhill County extension agent Wayne Roberts revealed 6,500 acres of fruit and nut orchards were destroyed out of a total inventory of 14,000 acres. The destruction far surpassed other weather disasters, including the orchard freezes of 1919, 1936, and 1955. The farmers who planted, tended, and harvested those orchards had never seen a wind like this one. Yamhill County was not alone. The grim losses played out across a six-county region of the Willamette Valley, with orchard damage pegged at more than $60 million, including 75 percent of the walnut trees in Marion County, 75 percent of the prune acreage in Washington County, and 50 percent of the prune orchard inventory in Clackamas County.

"What do the farmers do until new orchards are producing?" asked *Oregon Statesman* staff writer Floyd McKay in a story assessing Willamette Valley orchard damage published one week after the storm. "That's a million dollar question, and answers are being pondered right now in many farmhouses." Jay Greer, owner of a forty-acre nut and fruit orchard near Salem, Oregon, in Marion County recalled what it was like to come home to the storm damage from his day job supervising an inmate crew at the Oregon Fairview Home, an annex of the Salem-based Oregon State Hospital, which housed mental health patients. "My ten-acre filbert orchard was just destroyed, the trees were all mangled. And my five or six acres of prune trees were all tweaked, too. It was like a big hand pushed across my orchards from south to north." Newberg orchardist Bradley Smith watched the winds shred his walnut orchards from the window of his home. He lost about two-thirds of his filbert and walnut trees. "I was so hard hit," he lamented a year after the storm. "I could see no choice but to just walk away."

Farmers were faced with tough choices in the storm's aftermath. A walnut tree takes ten or more years to start producing, and thirty years to attain full production. Filbert and prune trees need six to eight years. Both the walnut and prune growers already faced challenging times before the storm hit. The Oregon walnut industry, promoted as the next big crop for Oregon in the early 1900s, had peaked twenty years before the storm and was steadily losing market share to California growers. The big freezes and blackline disease, a slow-growing virus found in the native black walnut rootstock used for grafting, was particularly hard on mature trees. The winds of the Columbus Day Storm were a final blow to the industry. The state's prune industry was falling on hard times, too. Prune production in 1962 had been forty-five thousand tons, the largest yield since 1959. But a large percentage of those plums fell to the ground and rotted because of a lack of pickers, low prices, and a shortage of processing plants. By 2014, prune production had dropped to 7,800 tons, which was still more than 50 percent of US production.

The storm, market forces, winter freezes, tree disease, and the pressure of population growth on farmland collectively closed the door on the once-robust walnut and prune industries in Oregon. Only filberts, more commonly known as hazelnuts, continued to see growth in production, climbing from three hundred tons in 1930 to fifteen thousand tons in 1980 and thirty-six thousand tons in 2014, making it the eighth-most valuable farm commodity in Oregon, which ranked number one in hazelnut production in the United States.

There is another crop and industry doing well these days in Oregon, one that was just taking root in the hearts and minds of two or three enterprising souls in 1962. The number of grape vineyards for wine production was fast approaching one thousand in 2014, placing grapes tenth among Oregon crop values, and making it Oregon's second most valuable fruit crop. The state is home to more than six hundred wineries, triple the number from only a decade ago, and the third most of any state in the nation. A wine industry that features a statewide economic impact of more than $3.3 billion a year didn't even exist when the Columbus Day Storm struck more than fifty years ago. But the storm, the Dundee Hills, and the Oregon wine industry are inextricably linked.

While orchardists in the north Willamette Valley were rubbing their foreheads and straining to see past the Columbus Day Storm damage into an uncertain future, two 1962 graduate school classmates in University of California at Davis' Department of Viticulture and Enology were pursuing their own radical vision that would lead to commercial grape-growing in the valley. One was Charles Coury, an Oregon native with a bachelor's degree in climatology from the University of California at Los Angeles. The other was David Lett, a fair-haired Utah farm boy bound for dental school until he took a life-changing trip through California's Napa Valley wine country. Their studies and their willingness to question conventional wisdom kept pointing them north to the orchards and farmlands of the north Willamette Valley, which they fancied as a prime location for growing cool-climate grape varieties, principally Pinot Noir, a deep-purple grape that is tricky to grow but produces a complex, light red wine with hints of cherry, berry, and earthy aromas that tell a story of where the grape is grown.

After graduating, Coury and Lett took their thirst for knowledge about grape-growing and wine-making on independent journeys to Europe in 1964. Lett spent time in Burgundy in northeast France and Coury visited Alsace on the German border, two wine-producing regions at latitudes similar to those of the north Willamette Valley. They came home, planted nursery stock in the Willamette Valley in 1965, and began their search for suitable land to start vineyards in Oregon. David and Diana Lett settled in the Dundee Hills in 1966 on an eight-acre prune orchard abandoned after the Columbus Day Storm, renaming it Eyrie Vineyards. Chuck and Shirley Coury established their vineyard and winery in the Forest Grove area north of the Dundee Hills. The Willamette Valley wine industry was born and, over time, Lett earned the nickname "Papa Pinot."

Another Oregon wine–pioneering couple, Dick and Kina Erath, grabbed up a fifty-acre walnut orchard in the Dundee Hills above Newberg. "Between the freeze in '55 and the Columbus Day Storm, it pretty much took them [the walnut trees] out," Erath, an Oakland, California, native and electronics engineer by trade, said of the orchard's condition when he bought it in 1968. The wine pioneers intent on growing the Pinot Noir grapes and other cold-weather varietals kept filtering into the Dundee Hills area, including Stanford University graduates Bill and Susan Sokol Blosser. In late 1970, they found an eighteen-acre prune orchard, on a hillside above Highway 99 West, just southwest of Dundee. It, too, was an orchard victimized by the Columbus Day Storm, covered in blackberry vines and vetch, the trees lying haphazardly on the ground, few still upright.

"It was as if the orchard had been waiting all those years for us to discover it," Susan Sokol Blosser recalled. "And the property that we bought. . . . How do you know what land is going to be good for growing grapes?" she asked. "Well there was no way of knowing because there were, you know, no producing vineyards at that point." The young couple plunked down $800 per acre, relying on a dosage of wishful dreams and intuition to justify their investment. "Knowing fruit trees had flourished there cinched our decision to buy the land," she said in her 2006 memoir, *At Home in the Vineyard: Cultivating a Winery, an Industry and a Life*. "It meant the hillsides were frost-free in early spring when those trees blossomed." It mirrored the time of year when grapevines leaf out and blossom.

When the Sokol Blossers arrived in the Dundee Hills, local farmers were still talking about the Columbus Day Storm like it was yesterday. One of those farmers, who predated the wine pioneers by several years, was Jim Maresh (pronounced Marsh), a Wisconsin native, retired commander in the US Navy Reserve, and a business analyst with Dun & Bradstreet in Portland, Oregon. Maresh, his wife, Loie, and their five children never intended to be a farm family. However Maresh fell in love with the rolling, red hills of Dundee, purchased a twenty-seven-acre farm covered with plum, cherry, and filbert trees, and moved his at-first-reluctant family from the city to the country in 1959. He didn't know much about farming, but he was a quick study, picking up advice from the older farmers in the Dundee Hills. And when some of his older neighbors grew too tired or too old to farm, he purchased their properties. With support from another neighbor, newcomer Dick Erath, Maresh pulled up his prune orchard and planted three acres of grapevines in 1970.

Today the Maresh estate encompasses 124 acres of vineyards featuring several varietals, include Pinot Noir and Chardonnay. Maresh, approaching ninety and still working the grape harvest in 2015, minced no words about the role the Columbus Day Storm played in the development of a world-class wine industry in Oregon, particularly the Dundee Hills he has called home for more than fifty-five years. "The only good thing that came out of that storm is it completely destroyed the Oregon prune industry," he told *Oregonian* reporter Katherine Cole in May of 2015. "And that was the best thing that ever happened to the Oregon wine industry."

The conversion of storm-battered prune orchards to vineyards extended beyond the Dundee Hills. Oregon wine pioneers Joe and Pat Campbell found another prune orchard abandoned after the Columbus Day Storm some twenty miles north of Dundee, near the northern edge of today's wine-growing country in the foothills of the Oregon coastal mountain range. Joe Campbell, thirty-five, was a Stanford Medical School graduate and emergency room doctor who filed as a conscientious objector during the Vietnam War. To avoid the draft, he took a physician's job on the Sioux Pine Ridge Indian Reservation for two years, and that's where their son, Adam, was born. He was three years old when his parents moved an old trailer onto the property in Oregon and began pursuit of their grape-growing, wine-making dream. In 1999, Adam Campbell became the owner and head winemaker at Elk Cove, one of several sons and daughters of Oregon wine pioneers carrying on the tradition their parents started.

I emailed the younger Campbell in early September 2015, the year marking the fifty-year anniversary of the Willamette Valley wine industry. I was just making sure Elk Cove Vineyards was sitting on an old prune orchard, torn apart by the Columbus Day Storm. "Funny. I just thought about the storm yesterday," the younger Campbell responded. "As we are approaching harvest, my Mom always cooks a prune tart from the fruit of one of the few remaining trees we have on the property, and it's interesting to think of the place covered in prunes, not grapes. So our place was one of the many prune orchards abandoned after the Columbus Day Storm. Seeing the trees gave my parents the hope that this would be good ground for grapes.... Since it was abandoned it was also very inexpensive—$1,100 per acre for 112 acres—another requirement for the early days of the industry."

Approaching Dundee on Highway 99 West from the southwest, the traffic moved at a snail's pace. The mile-long backup was but a slight annoyance on a picture-perfect Sunday afternoon in late September of 2015. There were blue skies, warm sun, and vineyards galore, many of them already stripped of their bounty thanks to a droughty, hot summer that had ripened the grapes, including the swollen, purple-black Pinot Noir, a month earlier than usual. One of the dozens of blue-and-white road signs sprinkled on the hillsides, directing wine country tourists to wineries and tasting rooms, pointed to the Sokol Blosser Winery.

Some forty-five years after the Sokol Blosser Volkswagen camper-bus came to rest in the Dundee Hills, the wine-pioneering family presides over an organic certified eighty-six-acre vineyard, a state-of-the art winery, and a low-slung, terraced tasting room, striking in its light, angles, and use of tight-knot cedar, Douglas-fir, and hickory to shape the exterior and interior walls, floors, and ceilings. The tasting room, which opened in July of 2013, is both elegant and earthy. A crowd of forty or so mostly millennials stood around the tasting bar, sat in gray overstuffed chairs near a large fireplace or gathered at tables on the veranda. Everyone was greeted by stunning views of the Willamette Valley to the south and the Cascade Range, including Mount Hood, to the east. They were all sipping flights or full glasses of wine that tended toward the Dundee Hills go-to wine—Pinot Noir.

I had arranged to meet Susan Sokol Blosser for an interview. She strolled into the tasting room largely unnoticed by the animated, wine-sipping crowd, dressed in a long-sleeved cotton shirt, tan jeans, dark blue vest, and baseball cap atop a salt-and-pepper head of hair. She was a petite woman approaching her seventy-first birthday, with an easy smile, smooth complexion, and dark eyebrows that arched as she pondered answers to my questions.

We stepped outside and sat at a wooden picnic table that afforded views of the original vineyard, surrounded by newer additions. I asked her about the link between the Columbus Day Storm and the birth of the Oregon wine industry. She chose her words carefully, reluctant to cast in a favorable light a storm that claimed the lives of two men and a teenage boy and girl from the neighboring towns of McMinnville and Dayton. "The birth of the industry was really a perfect storm of events, to use the term in a good way," she said. Yes, the storm brought land on the market, but that land would have lain fallow, if not for the risk-taking, back-to-the-land entrepreneurs short on cash but rich with vision who staked a claim and followed their dreams. And, she added, don't overlook Oregon's landmark growth management law, passed by the state legislature in 1972 and signed into law by Governor Tom McCall—the former KGW-TV news reporter working with Capell in Portland when the storm struck. It provided the framework for Yamhill County elected officials—urged on by the wine pioneers and established hillside farmers, including Maresh—to preserve the Dundee Hills for agriculture, not allowing developers to build housing subdivisions, trailer parks, or, worse yet, an asphalt plant once proposed for the area.

As we talked, a double-decker red bus filled with wine tourists from Portland, Oregon, pulled into the winery parking lot, paused, then turned around to leave. "They're probably looking for the Domaine Drouhin winery," Sokol Blosser remarked, a bemused look on her face. She was referring to a neighboring vineyard owned by the Drouhin family of Burgundy, France, which has been making world-famous wines since 1880. The family expanded into the Dundee Hills in 1988, an affirmation that the north Willamette Valley is exceptional Pinot Noir country, too.

What about the future? What are the challenges ahead for her son and daughter, Alex and Alison Sokol Blosser, copresidents of the family business? With a long, hot summer still fresh in her memory, Susan Sokol Blosser wondered whether her children face a vineyard do-over in the years ahead, a transition to grape varietals suited to warmer weather caused by climate change.

And what about today's or tomorrow's vineyards—Would they survive a Columbus Day Storm better than the fruit and nut orchards did more than fifty years ago? The wine pioneer swiveled her neck and gazed at the vines as if looking for an intruder that may or may not exist. With their ground-hugging profile, the vines look far less vulnerable to punishing winds than fruit and nut trees were in 1962. "I sure hope so," she said.

10

Bridgetown under Siege

Busy but not in a hurry, surrounded by natural beauty but dissected by a grossly polluted Willamette River, the city of Portland, Oregon, on October 12, 1962, was still a coarse, low-slung, misty frontier town, absent signs of the smart-growth future it would have, a future filled with coffeehouses, microbreweries, two hundred city parks of all sizes, and one of the largest independent bookstores in the country, Powell's City of Books.

Hemmed in by the river to the east and the Tualatin Mountains to the west, Portland's population was spreading eastward—aided by nine bridges that spanned the river—beneath towering Mount Hood, a jagged southeastward volcano that stands 11,245 feet tall, ever watchful over the city from some fifty miles away. It's the most climbed mountain in the country, second in the world to Mount Fuji. The Morrison Street Bridge, the first of those bridges and the oldest west of the Mississippi River, wasn't completed until 1887. It marked the beginning of the end for a ferry system that started in the mid-nineteenth century with the *Black Maria*, a paddle-wheel ferry powered by a disciplined mule on a treadmill and guided across the river by a submerged cable. The bridge toll was five cents for people, sheep, and hogs, twice that for horses and cattle.

On the night Portland was ambushed by the Columbus Day Storm, those bridges became dangerous, windblown places. The wind gusted to 116 miles per hour on the Morrison Street Bridge at 6:14 p.m. Then the wind gauge blew away. At the peak of the storm, only four of the nine bridges spanning the Willamette in Portland were open. Police closed the St. Johns Bridge, a 1931 suspension bridge, when the winds triggered a fifteen-foot vertical sway of the bridge deck.

The day before the Columbus Day Storm, a windstorm had battered the Oregon coast and dumped two-thirds of an inch of rain on Portland.

But that early fall storm was all but forgotten Friday as "Rose City" residents enjoyed a blue-sky day, one of only sixty-eight they see on average each year. Residents of Oregon's largest city were mostly oblivious to the impending doom, despite an ominous forecast issued by the US Weather Bureau's Portland office mid-morning Friday. Another storm front was approaching Western Oregon and Washington. "Expect wind gusts to 69 miles per hour Friday night," the forecasters predicted.

The wind began to stiffen as rush-hour traffic filled the city streets and bridges. Meteorologist Jack Capell had issued his warning of destructive record winds on KGW radio's 5:15 p.m. weather report, just minutes before the storm struck. Some Portlanders heard Capell's weather warning. Many didn't. Missing from the radio airwaves that fateful Friday night was KISN radio's Operation Airwatch, the morning and evening commuter traffic report delivered over the skies of the Portland metropolitan area from Aerocar N103D, a part airplane–part car contraption built by Longview, Washington, aeronautical engineer Moulton Taylor.

The airwatch pilot Friday night was Ruth Wikander, a flight instructor and manager of Wik's Air Service, based at Hillsboro Airport some seven miles west of Portland. Shortly after takeoff, the winds began to blow her light flying machine sideways. She realized she would never make it to Portland. Wikander, forty-three, aimed the Aerocar into the teeth of the wind and hoped for the best. "I was flying backwards at full throttle," said Wikander, recalling the hour she battled the buffeting winds before they died down enough for her to land.

Wikander's was one of several dramas that played out in the skies over the Pacific Northwest October 12, 1962. Planes on the ground and in airport hangars across the region fared even worse. Some 175 planes were tossed around like toys and damaged or destroyed in the Portland, Oregon, and southwest Washington area. Two dozen airplane hangars and 226 planes at thirteen airports were damaged or destroyed along the path of the storm in Western Oregon. The Oregon State Board of Aeronautics later estimated the airplane and hangar damage at nearly $13 million.

Every time Wikander tried to climb above 1,000 feet altitude, the hybrid plane, with a 143-horsepower engine and top speed of 110 miles per hour, was swept backward. "When we dropped down lower, the air was so rough and turbulent it was all I could do to keep it under control," Wikander told an *Oregonian* reporter the day after her frightening ride.

One minute Wikander and her passenger, company employee John Jacob, were plunging toward the ground in the all-white novelty aircraft, which looked like a stubby Fiat attached to a Cessna airplane frame. The next minute the Aerocar was climbing the skies at a 45-degree angle.

Wikander, a competitive, accomplished aviator with unruly tawny hair and a self-assured grin, was a proud member of the Ninety-Nines, Inc., an international organization of women pilots inspired by pioneer female aviator Amelia Earhart. Wikander was a woman who never married, moving freely through an aviation world still dominated by men, adorned in her khaki pants, T-shirts, and leather aviator jacket. She had lost her older brother, Guilford Wikander, just five months before the Columbus Day Storm when the small plane he was piloting crashed near Bend, Oregon. He was a World War II pilot and president of Wik's Air Service at the time of his death. He must have been on his sister's determined mind as she struggled to keep the floaty, hybrid plane aloft. "Once the storm dragged us down 1,000 feet," she said of the traffic watch mission gone awry. "I managed to pull out about 150 feet from the ground."

As Wikander fought to get her part-plane, part-automobile on the ground, Portland television meteorologist Jack Capell searched for an unfettered route back to the KGW office in downtown Portland. The news team was operating with auxiliary power and anxiously awaiting Capell's arrival at the station with an explanation for this sudden, hellish windstorm striking Portland. He traveled across the white-capped Willamette River on the Steel Bridge, disobeying a police officer's order not to use the bridge.

Almost out of fuel and still in the storm's grip, Wikander knew she had to try to land. "We came straight down, like an elevator, with nearly full throttle," she told the *Oregonian* reporter. She landed in the grass between runways. "Once on the ground, it took full throttle to hold the Aerocar against the wind until a half dozen men could grab the wings. Then we bailed out." The dual airplane-car, red hearts painted on the top and bottom of each wing, wasn't damaged in the air. But once on the ground the wind's force popped out the windshield. The wind tipped the dual-purpose craft on its side, denting a wing.

Airplane carnage was widespread at the Hillsboro Airport, and at several other airports in the storm's path. About sixty planes and six hangars were destroyed or damaged at the Hillsboro Airport. The wind had

Small airplanes such as these two at the Kelso Airport in Kelso, Washington, were tossed around like toys by the storm's winds. Courtesy of the Cowlitz County Historical Museum, Kelso, Washington.

no trouble lifting flimsy, light airport hangars into the air, sending them sailing and cartwheeling through the air before they crashed into pieces on the ground. Exposed planes were picked up by the winds and propelled in awkward flights that ended badly at municipal airports all over the Pacific Northwest.

At Portland International Airport, a new hangar was tossed off its foundation and flattened, damaging five planes. The wind tore a twin-engine Lockheed from its mooring at the airport, upending and blowing it across the north runway into a boundary fence. Several planes were scoured by blowing sand at the peak of the storm. The wind lifted the roof from the new hangar-lounge built by Pacific Power & Light Company at the Portland–Troutdale Airport. Falling beams damaged two twin-engine planes used by utility executives. Some thirty-six planes and half a dozen hangars valued at more than $1.5 million took a beating at the airport.

At Pearson Field in Vancouver, Washington, just across the Columbia River from Portland, Clyde Wells tried to save his airplane from the first gusts of the storm. A nearby hangar exploded, sending debris into a gathering of men, including Wells. Bystander Kenneth Poe said it was a miracle no one was killed. "We all dived for shelter, but pieces of the roof and sheet metal sides blew all around us," he said. Wells was seriously injured and had to be dragged clear of the debris and transported to Vancouver Memorial Hospital. About sixty of the seventy-five planes and all of the hangars were destroyed or damaged at the airfield, which is the oldest in the Pacific Northwest. Some planes were hurled more than a thousand feet across a road, and others lifted off the ground and landed on crumpled hangars and planes.

As the winds picked up pace and velocity, longshore workers Francis Murnane and Stanley Zagorski left the ILWU hiring hall on NW Glisan Street and headed to work at Terminal 1 on the Willamette River just north of downtown Portland. This is where the nation's longest north-flowing river curves slightly to the northwest before meeting the Columbia River. The shipping docks, wharves, and terminals, many of them dedicated to loading cargo ships with grain, were scenes of chaos and destruction.

Murnane, president of the International Longshore and Warehouse Union Local 8 and an avid city historic preservationist, described the scene near Terminal 1: "We saw the metal sign blow off the roof of the Noon Bag Company. Traffic signals on NW Front Avenue were dancing wildly in the wind. We saw heavy planks peeled off a stack of lumber as though they were playing cards. Glass and skylights were tumbling from the warehouses on the dock. The metal doors were ripped loose and dangled in a grotesque manner."

Despite the danger and awful power of the storm, Murnane and Zagorski were drawn to the edge of the dock. They soaked up the stormy sights. "The Willamette River was very rough and choppy and we stared in amazement," Murnane said. "We saw half the roof ripped off the Albina Dock across the river, and land in the Union Pacific [railroad] yard. The front of the Northwest grain dock was smashed like kindling wood. Bits of glass and wood were flying in all directions. But danger has a strange fascination, and we continued to watch."

They watched the wind pry the cargo liner *M. S. Washington* from its mooring at the Crown Mill dock. The huge ship was swept across the river, and three tugs maneuvered desperately to keep the 565-foot-long vessel from crashing into the river's east-side bank. "Three huge hatch tents had been ripped from the hatches and were dangling at the [ship's] side like big serpents. Lines were dangling and constituted a grave danger to the crew," said Murnane, a slingman whose job it was to stand on the wharf and attach the hoisting tackle to the dockside cargo headed to the ship.

South of downtown, at the Zidell dock, where navy and merchant marine ships were dismantled after World War II, a 350-foot-long navy vessel broke loose and crashed into the oldest lift bridge in the world, Hawthorne Bridge. Originally built in 1891, then rebuilt in 1910, the bridge was closed Friday night because of storm damage to the railing and walkway, but three of four lanes were reopened the following day. Damage to yacht clubs and houseboats along the Willamette and Columbia Rivers was severe. At the Totem Pole Marina on the Columbia River, near the interstate bridge connecting Oregon to Washington, seventeen recreational boats on display were blown over by the winds, one landing on a parked car, and outboard motors were strewn about the marina showroom like toys. John E. Lenz, fifty, the marina manager and recently elected president of the Oregon Marine Trade Association, tried to batten down the marina hatches, but suffered a fatal heart attack trying.

In addition to John Lenz, there were three other freakish storm fatalities in the Portland area born of the fierce, unprecedented winds. Two-year-old Michael Gensel died when he was struck by a falling tree in the backyard of his Portland home. The toddler had followed his mother into the yard without her knowledge. In north Portland, farmers from all over Oregon gathered with their livestock Friday at a sprawling complex of exhibition halls and barns known as the Pacific International Livestock Exhibition. Saturday marked the opening of the exhibition center's annual livestock show. The winds clawed at the old wooden buildings, dislodging roofing and siding that sailed through the air. Leo J. Buyseries, a prominent farmer from Rickreall near Salem, Oregon, was fatally injured after being struck by wooden debris from a collapsed roof. The winds blew the horse barn apart, sending horses running amok on the exhibition grounds. The exhibition did open Saturday, despite Friday night's tragic events. Twenty years before the storm, the barns and corrals at the exhibition center had been used to confine more than 3,650 Japanese Americans rounded up on the West Coast for internment during World War II. They sat for five months in stalls with plank floors hastily installed over livestock-soiled hay and sawdust, awaiting assignment to more permanent internment camps in California, Idaho, and Wyoming.

A fourth storm fatality in Portland occurred at one of the largest shopping malls in the country, the hundred-store Lloyd Center in northeast Portland, which had opened to much fanfare in 1960. Harold E. Morrison, thirty-seven, a district sales representative for Philco-Ford Corporation, stepped over a parking garage railing in the dark and plunged seventeen feet to his death. His body was discovered around midnight. The fatal fall apparently occurred sometime after 7 p.m., when the Portland area suffered a widespread power outage that took several days to completely restore.

Storm-related injuries sent some 161 persons to Portland hospitals, where staff worked through the night on auxiliary power. Of those injured, fifty-five patients were admitted with injuries from flying glass and falling trees and limbs. The storm chewed up suburbs and towns in the greater Portland metropolitan area as well. In Oregon City to the east, two steel trusses weighing five tons each tore loose from a concrete foundation, sailed three hundred feet through the air, and smashed into two homes, demolishing one of them. In the heavily wooded Portland suburb of Lake Oswego, 70 percent of the four thousand homes were damaged by falling trees during the storm. In the nearby town of Gresham, the winds flattened a bowling alley and restaurant. An ensuing fire consumed the remains of Eastmont Lanes, which had opened just a year before the storm hit.

The winds held Albert and Hazel Capen hostage in their Portland-area suburban home, a solar-powered labor of love that had taken six years to build. The winds unraveled their work in less than an hour. Trapped inside the home by flying glass, there was no chance for them to escape. They dropped to the kitchen floor, joined by their black cat, Shmoo, and watched in horror as the roof lifted off their house and two thirty-foot-long walls, peppered with shingles, tree limbs, and broken glass, collapsed. They moved into the utility room from the kitchen when they couldn't keep the kitchen door wedged shut. Oddly enough, the telephone in the utility room still worked. Capen, fifty-five, called his mother and told her he didn't think they would survive the storm's assault. "We thought we were having a hurricane," he said the next day. "We were blowing up."

The assault on the home continued. A Roman brick fireplace was sheared off at the top. Four steel pipes that anchored their demolished patio were tossed in the yard like sticks. The couple's possession—chairs, lamps, rugs, a record player, tables, a television set, and china closet—were shattered and strewn about the exposed interior of the L-shaped home.

City parks and the wooded neighborhoods of Portland lost up to sixteen thousand trees that were knocked to the ground or shattered by the winds. Some of them predated the settlement of "Stumptown," one of Portland's many nicknames. Ted Buehner, a weather warning meteorologist with the National Weather Service in Seattle, was a six-year-old boy living on Highland Drive just south of Forest Park and a mile west of the Portland Zoo the night the wind hit. His childhood remembrances of the storm included a hurried trip to the grocery store with his mother to buy emergency provisions. Etched in his memory is the sight of yellowish-green clouds to the west before the winds struck. Those colors suggested a storm approaching from about twenty-five miles away. The clouds were ripping through the sky at 40 miles an hour or more. The Buehners barely had time to get home before the wind arrived.

Gathered in the kitchen with his mother and two sisters, Buehner and his family played Bingo by candlelight and ate twice-baked potatoes for dinner. Buehner remembers the winds howling, whistling, pausing, and howling again as dusk turned to night. "Trees were crashing and limbs were hitting our roof—we lost some gutters and some shingles off the roof," he said, recalling the storm fifty-one years later while on duty at the NWS station in Seattle. "The spookiest thing for all the kids was hearing trees crashing and thudding to the ground in the dark." In the morning, Buehner counted 160 trees blown over on his neighborhood street, which was without electricity for ten days.

The storm was a life-changing experience for Buehner, causing the young boy to wonder what caused such destructive winds. By age ten, he had his own home weather station. He built a weather station at his high school in Lake Oswego. As a senior, he taught basic weather concepts to sophomores. He attended Oregon State University, earned a degree in atmospheric sciences, and joined the National Weather Service in 1978. He credits the Columbus Day Storm for arousing his lifelong curiosity in severe weather and propelling him to a career as a meteorologist.

In 5,100-acre Forest Park—one of the largest city parks in the nation— the winds laid waste to hemlock, fir, pine, alder, spruce, dogwood, and willow trees, leaving in its wake two million board feet of fallen timber, blocking a network of trails first envisioned in 1903 when the Brookline, Massachusetts–based Olmsted Brothers landscape architectural firm developed a plan for Portland parks. It took years to patch together the public ownership, but the park was finally dedicated for public use in 1948. Perched along the hillside running eight miles parallel to the Willamette River, the park reaches an elevation of 1,100 feet. It is an urban oasis of greenery, wildlife, and stunning vistas of the river, downtown Portland, and distant snow-capped peaks in the Cascade Range. After the windstorm, it took a crew of twenty workers about 7,600 hours to clear the roads and seventy miles of trail.

Storm damage proved to be a blessing in disguise for the Pittock Mansion, a four-story, twenty-two-room home built in 1914 by *Oregonian* newspaper publisher, Portland pioneer, and business magnate Henry Lewis Pittock. The sixteen-thousand-square-foot edifice, built with chunks of Tenino sandstone, some weighing up to thirty-five tons, had sat empty since the last of the Pittock family moved out in 1958. The storm hit the French chateau-style mansion hard, breaking windows and peeling off about one-third of the tile roof. The Pittock family was poised to demolish the mansion and subdivide the forty-six-acre spread into a high-end housing development. But city historic preservationists, including longshoreman and community activist Francis Murnane, prevailed on the city to purchase the home and property for $225,000 in 1964 and commit to preserving the mansion and grounds as a city park.

The post-storm fight to save the Pittock Mansion was just one of several that Murnane engaged in as labor leader, member of the Portland Arts Commission and Oregon Port Commission, and Multnomah County planning commissioner during the 1950s and 1960s. Murnane, the son of immigrant Irish parents, was born in Boston, but moved to Portland as a young child of the Depression. His career plans morphed from priest to lawyer to union labor activist, and he joined the Progressive Party as a young man. He was a husky bachelor with thick black hair and a head-up, chin-up, confident stride. He later turned away from radical labor-organizing politics, presiding over a longshoreman union local reluctant to accept black members.

In 1958, joined by legendary Oregon journalist and populist historian Stewart Holbrook, Murnane led the charge in downtown Portland to restore the twenty bronze drinking-water fountains donated to the city in 1912 by Simon Benson. Fifty years before the storm, Benson had been motivated to install the so-called Benson bubblers by more than largesse. He wanted an ample water supply available to millworkers and loggers,

Objects in the path of collapsed brick and other storm debris in downtown Portland, including this car, were crushed beyond repair. Courtesy of the Oregon State Historical Society.

many under his employment. He thought too many of his employees were spending idle time in saloons and bars, drinking beer to quench their thirst.

Murnane was also instrumental in efforts to save and restore the Skidmore Fountain near the west end of Burnside Bridge in Portland's Old Town. Built in 1888 for the benefit of horses and humans alike, the fountain is Portland's oldest piece of public art. As the winds subsided the night of the storm, Murnane was drawn to the streets of downtown Portland.

"I had to see what was happening. There might be injured people lying in the streets and needing help," he said. "I wondered about the trees in the parks, and the fountains and memorials that I had worked so hard to have restored. It was an eerie experience walking along darkened and deserted streets."

He walked east on SW Columbia Street to the South Park Blocks: "The devastation was appalling. Beautiful trees which had graced the parks for generations were blown down and splintered in all directions. It was like a nightmare. It didn't seem that such a horrible thing was really happening. . . . The beautiful steeple of the First Congregational Church was badly damaged and brought to mind the bombed churches in London [in World War II]."

All things considered, city landmarks and memorials fared well. A huge tree fell just inches from the statute of Abraham Lincoln, and the figure of Theodore Roosevelt astride his steed was unscathed. The Shemanski and Benson Memorial Fountains were spared as well. At Plaza Park on SW Salmon Street, trees littered the ground, but the Spanish American soldier of war remained on his pedestal, rifle in hand, marching on the enemy. "Walking about the downtown streets was dangerous, yet strangely fascinating," Murnane recalled. "Huge plate glass windows were shattered and the pieces lying on the sidewalks or streets. Here and there, mannequins, attired in expensive clothing, were lying about like the dead after a bombing. Police and private guards were on duty before windowless jewelry stores. Electric signs were ripped from buildings. I proceeded west on SW Morrison Street. No one else was about."

11
Life Turns on a Dime

As darkness descended on the Portland area, heightened by widespread power outages, the windstorm raced across the Columbia River, the 1,200-mile-long force of nature that flows through seven western states and British Columbia, marking the border between Oregon to the south and Washington to the north. The hundreds of houseboats, or floating homes, clustered along the banks of the Columbia and Willamette Rivers near Portland were especially hard hit. Many were ripped apart by the winds or pushed from their moorings on the Oregon side of the mile-wide Columbia River, losing windows, doors, and roofs before coming to rest on the Washington shoreline near Vancouver, Washington.

The storm claimed three more lives in Vancouver, including Wayne C. Goheen, seventy-seven. A cluster of poplar trees in his front yard fell, caving in the corner of the living room where he sat, watching the winds assault his property. His relatives said Goheen had planted the trees forty years before the storm, when he moved to Vancouver. They most likely were planted as a source of shade, and as a windbreak that fatally failed to hold the line. Goheen's death was another sad reminder of how wind-toppled trees were the force multiplier that made the storm so deadly, and so damaging.

The storm kept moving north through southwest Washington, closing in on the population centers of Puget Sound. At the top of Mount Pleasant in rural Kelso, Donald L. Ames, thirty-five, fought his way through the wind from his house to his aluminum barn to rescue a calf, but the storm intervened. A blast of wind toppled the barn, sending timbers and metal crashing down on Ames. Dazed and bleeding heavily from a scalp wound, Ames staggered back to the house where he received first aid from his cousin and neighbor, Bob Boone.

Boone set out on foot in the howling winds, headed down Mount Pleasant Road in search of help. The road was a 2.5-mile obstacle course of fallen trees and limbs, impassable by vehicle. Boone encountered another neighbor, Phil Jacobs, who had a car at the bottom of the mountain road. The two men hiked to the car and drove to Cowlitz General Hospital, where they secured a stretcher and assistance from hospital board member Donald Van Brunt. The three men headed up the hill with the stretcher. The road had been partially cleared by residents, who also loaned the rescuers their cars. Joined by a civil defense crew that was checking the isolated area, they reached the Ames residence at 12:30 a.m., more than six hours after the accident. By then the road had been cleared enough for a truck to transport Ames to the hospital. He was treated and reported in satisfactory condition later Saturday. It's possible he could have bled to death without the determined aid of neighbors and strangers.

Fifteen miles southeast of Olympia, on the outskirts of the small prairie town of Yelm and in the shadow of a towering Mount Rainier, Wilbur Roy Archibald—his friends called him Roy—had just finished pouring the concrete foundation Friday for a home he was building for his family on sixty-eight acres off Bald Hills Road. Learning on his transistor radio about the fast-approaching storm, he drove home, just a few minutes away, to pick up his wife, Carol, to help him secure a tarp over the foundation to protect it from any pending rain. Three of Archibald's five children were at home—Bruce, twelve; Susan, seven; and Tim, five. The two other siblings, Nancy, fourteen, and Shelley, ten, were at friends' houses. The parents felt comfortable leaving Bruce in charge of his younger brother and sister. After all, they would be gone for only a few minutes. It appears the couple completed their chore. They were northbound, headed back home, when three large Douglas-fir trees along Bald Hills Road were caught in a gust of wind sweeping across the prairie and toppled onto the road. One crushed their station wagon and the other two pinned the car forward and back.

State Patrol trooper Emmett Smith was working the Olympia area that Friday night. He was alerted early that evening by state patrol headquarters in Olympia that a punishing windstorm was approaching from the south. He was instructed to contact Olympia downtown businesses and warn them of potential damage to their buildings. He also called home to warn his family of the storm about to strike. "I'd no sooner hung up the phone than I got word of a fatal accident in Yelm," Smith told me twenty-five years after the storm. It took Smith at least two hours to reach the scene of the Archibald's accident, a trip that normally would take about thirty minutes.

"The wind must have been blowing 100 miles per hour," Smith recalled, although official wind gusts in the Olympia area were closer to 80 miles per hour. On his way, he encountered a pickup and camper blown over and pressed against the railing of a freeway overpass bridge high above Capitol Lake by the State Capitol Campus. A man got out of the pickup and Smith grabbed him just as a gust of wind was about to blow him off the bridge. The state trooper then checked the passengers for injuries. There were none. He transported them to the Tumwater police station and continued his drive to Yelm, but was continually delayed by fallen trees across the road.

Oddly enough, though, people with chain saws always seemed to turn up right away to clear the road. By the time he reached the accident scene, a Thurston County sheriff's deputy and Yelm police officer were there. There was nothing they could do for the couple. Their injuries were fatal: Wilbur R. Archibald, forty-one, a broken neck, and Carol Archibald, thirty-seven, a fractured skull, according to Thurston County coroner Hollis Fultz. In a matter of seconds, a close-knit family had unraveled. A random, unfathomable disappearing act transformed five children into orphans, forever altering the trajectory of their lives.

In the summer of 2015, Susan Archibald shared with me memories of her parents, and what happened in the hours after the horrific accident. The three younger children were picked up by someone—Susan couldn't remember who—and taken to the home of nearby family friends, Leo and Marion Lefebvre. It wasn't until the next morning that Marion, terribly upset and crying, delivered the bad news to Bruce, Susan, and Tim. How Nancy and Shelley learned of their parents' deaths is less clear, but Susan said she heard them say later the life-altering message was not delivered to them in a very sensitive way. "It was such a shock," Susan Archibald said. "I remember thinking this seems like something that would happen to somebody else. I pretty much just shut down. It took me a long time to even talk about it."

Newspaper accounts of the Archibald parents and their fatal accident were cursory at best, delivering only occupation, age, and cause of death. They were treated as nothing more than storm statistics in the aftermath of the storm. As a young adult, circa 1970, I lived in the same house with Nancy Archibald while attending Western Washington University in Bellingham. She never talked about her parents, or what had happened to them. But she always seemed to me a bit lost, searching for her center, struggling to find her way. I knew Bruce, too, a former Olympia Brewing Company brewery worker, who died in 2003 at age fifty-three. In the summer of 2015, I tracked down Susan Archibald, a personal care products entrepreneur and year-round resident of Green Valley, Arizona. Through emails and phone interviews, she was the first of the surviving siblings willing to share memories of her parents.

Wilbur R. Archibald was raised in Hayward, California, and Carol (her maiden name was Lind), grew up in Tacoma. The two met while Roy was serving in the army, stationed at Fort Lewis near Tacoma during World War II. He later attended Pacific Lutheran University, a small private college near Tacoma, and earned a degree in geology. They made a striking couple: they were both redheads and good-looking. All five of the children were redheads and freckle-faced, too. On family outings to the ocean beach or to a family cabin on Whidbey Island in Puget Sound, the family of seven redheads attracted lingering stares from strangers. They were a sight to behold. Roy Archibald was a building contractor who owned his own company—W.R. Construction. The family had moved from Tacoma to Yelm in 1961, and the parents were already active in the community. He was a member of the Yelm Lions Club and she was active in 4-H and PTA. They even purchased a family plot in the Yelm cemetery, symbolic of their plan to sink some roots, raise their children, and live out their lives in rural Yelm.

The hillside east of downtown Olympia was a modest blue-collar neighborhood in 1962, overlooking an industrial waterfront of sawmills and plywood plants that hugged the Budd Inlet shoreline at the southernmost point of Puget Sound. This is the place where the Pacific Ocean, the birthplace of all the mighty Pacific Northwest windstorms, comes to rest. Stately Douglas-fir trees sat singly or in small clusters throughout the Eastside neighborhood, including in the front yard of a modest, two-bedroom bungalow owned by Karl and Carlene Pohl.

Carlene Pohl, twenty-four, a nurse's aide short in stature but long in spunk and spontaneity, was home alone the night of the storm, standing at the ironing board in her kitchen as the winds sang a shrieking song that set the neighborhood trees to dancing wildly. Her husband, whom she married in 1959 just ten days after they met, was working the swing shift at the nearby Georgia-Pacific plywood plant. A neighbor called Pohl and asked her to come over for a cup of coffee. Pohl obliged. The pile of clothes to iron could wait. Besides, the winds were too frightening to weather alone. She walked out the door, crossed the street, and entered her neighbor's home through the back door and offered a cursory greeting. Then she went to the living room and pulled back the curtains, remarking what a terrible storm it was.

"Looking out to the home I had just left, I saw a fir tree lying corner-to-corner across my home," Pohl told a reporter the next day. Where was the ironing board? Completely crushed and pushed through the floor in the spot she had been standing less than a minute earlier. "I would have been crushed," she said to me in a matter-of-fact tone more than fifty years after the death-defying event. The dissected home was a total loss, but the Pohls had just renewed their homeowner's insurance policy a few days earlier.

They used the insurance money and an FHA loan to buy seven and a half acres in rural Thurston County and build a new home. Her husband had always wanted to be a farmer, so they started raising pigs. The Pohls also cut down all the trees around their new home to make it safe from windstorms. Over the years, the farm grew to twenty-five acres and included beef cattle. "We just moved forward after the storm," Pohl said. Actually, Pohl hadn't liked the city house that much: It was where her husband had also lived with his first wife. "The storm was a blessing—a chance for a fresh start," she told me, allowing a wry grin to spread across her face. Her husband, who battled a series of serious health problems, died on October 16, 2012, four days after the fifty-year anniversary of the life-changing storm. Carlene Pohl is as spirited and forward-thinking as ever, still living on the now-dormant farm, and enjoying a life of travel and adventure, a life that was so close to being snuffed out by the Columbus Day Storm.

The storm came to life for me in Lacey, Washington, near the capital city of Olympia. The undefeated North Thurston Rams were prepared to take on their main rival, the Curtis Vikings in a Seamount League football game. I gathered in the grandstand with friends. With kickoff just minutes away, the wind morphed from a gentle breeze into a steady gale from the south. The field lights flickered and swayed in the wind. A state trooper cloaked in a cheap plastic raincoat and armed with a megaphone strode onto the field. He said a destructive windstorm was about to strike the Olympia area and urged everyone to head home immediately. The grandstand emptied and the crowd scattered in the wind. I was fourteen years old and knew next to nothing about severe windstorms.

My mother, father, two younger sisters, and I gathered at the home of Bill and Charlotte Kilde and their two daughters, Sandy and Sue. They lived in Thompson Place, a 1950s housing development carved out of a stand of second-growth Douglas-fir trees near Olympia, Washington. Despite the semi-wooded setting, my parents thought we would be safer there, compared to our own home set amid a dense stand of trees on a large rural lot a few miles away. Almost inseparable at the time, the nine of us were determined to ride out the storm together, gathered in the south-facing family room, which was a recent addition to the home and the room most vulnerable to the freakish southerly winds. We sat together, awestruck and frightened, and watched the storm unfold.

We were joined by Lee Bensley, owner of Lee's Steakhouse on nearby Martin Way, which was the main north–south road between Olympia and Seattle before the extension of Interstate 5. The popular roadside diner was home to one of the first Kentucky Fried Chicken franchises in the state of Washington, and had proudly played host to Colonel Harland Sanders not too long before the storm struck. "I was cooking oven fried chicken for dinner the night of the storm and remember worrying it wouldn't be as good as Colonel Sanders," Charlotte Kilde said.

Transfixed by the fury and freight-train roar of the gusting wind, we witnessed a sixty-year-old Douglas-fir tree smash the Kilde's garage roof and patio, a few feet from where we sat. Another tree crashed into the roof of a neighboring home. The backyard wooden fence peeled apart and sailed away in the wind. "We could have had wind-blown glass from the family room window in our faces, but we were mesmerized by the storm," said Kilde, a retired nurse with brilliant blue eyes and black hair turned silver, who still lived in the Thompson Place home more than fifty years after the storm.

After the winds finally died down, my father and I climbed into our old pickup truck and headed for home to check for wind damage, leaving my

The author rode out the storm in this storm-damaged, Olympia–area home. Courtesy of the
Olympia Tumwater Foundation.

mother and two sisters at the Kildes'. At least the Kildes still had electricity,
we figured. Who knew what we would find at our house? Traveling west
on Martin Way with very little traffic, we saw roadside signs tossed about
and power lines dangling from their poles. A sudden, final gust of wind
assaulted the vehicle and almost stopped us in our tracks. It was approaching 11 p.m., and I conjured up frightening images of our recently built,
rambler-style home in the woods, smashed to pieces by fallen trees.

Inching up our long dirt driveway to the crest of the hill where our
house stood in the trees, an eerie silence enveloped us. The landscape
was tattered and torn by a wind that had left us behind and continued its
northern pursuit. I dreaded what we would see. In the darkness I could see
fallen trees and branches strewn about the lawn and driveway, but much to
my relief, the house had been spared.

There was so much random tragedy and blinding luck along the path of the storm, lives turning on a dime. A station wagon traveling a second or two faster or slower on Bald Hills Road would have spared the lives of Wilbur and Carol Archibald, and kept the family intact. If Carlene Pohl had opted to stay home and finish her ironing, she would have been crushed by the tree that sliced through her house. Without the recent patio roof addition at the Kilde home, the Douglas-fir would have smashed into the family room where nine misplaced storm watchers, including me, could have been storm fatalities or among the seriously injured. And so it was on this windy trail of hits and misses, close calls, and sudden death.

On the fifty-third anniversary of the Columbus Day Storm, I paid a visit to the Yelm Public Cemetery, a well-groomed, rural burial ground that has served as the final resting place for the citizens of Yelm and the surrounding area since 1881. I was looking for the graves of Roy and Carol Archibald. I wanted to pay my respects to two of the dozens of people who died along the path of the storm. I rang the doorbell at the cemetery caretaker's home and office, hoping for directions to the Archibald gravesite. No one came to the door, so I slipped back into my small pickup truck and eased through the cemetery's black iron gate, determined to find their markers without any help. The cemetery, which encompasses a grassy knoll dotted with mature Douglas-fir, maple, and Garry oak trees, is only a few acres in size, and I had plenty of time on my hands, unlike the first time I tried to find their graves two months earlier during an unsuccessful cemetery drive-through just a few minutes before the gate closed for the night.

A slight breeze washed across the grounds, past the older section of the cemetery that includes many members of the Longmire pioneer family and its descendants, including James Longmire (1820–1897), who was in the first wagon train to cross the Naches Pass across the Cascade Range in 1853. He was a Mount Rainier guide and trailblazer and made the third documented ascent of the mountain in 1883. I zigzagged my way through the cemetery, angling back toward my truck, which was parked near the northeast corner. Closing in on a complete canvass of the burial grounds, I suddenly came upon the Archibalds' burial site, just fifty feet from my truck, their granite, low-profile headstones flanked by two ten-foot-tall juniper shrubs and a towering shore pine tree, with its thick scaly bark, multistemmed asymmetrical canopy, and small prickly cones, some closed and some that had opened years ago, still clinging to the tree branches. It was as if the graves were cloaked in anonymity in a purposeful way: no signs of visitation, flags, flowers, or grass freshly trampled by a visitor's footsteps, other than mine.

The inscription on the headstone is to the point and makes no mention of the Columbus Day Storm, or even the date of their deaths: "Carol Jean Archibald, June 1925 to October 1962" and "Wilbur Roy Archibald, August 1921 to October 1962." Their headstones have been joined by one marking their oldest son, Bruce Elliot Archibald, who was born on March 17, 1950, and died on May 2, 2003. Also laid to rest in the family plot are the paternal grandparents, Frank John Archibald, who was born on June 1, 1897, and died October 13, 1977, just one day after the twenty-five-year anniversary of his son's death, and Elsie Lyle Archibald, who was born March 8, 1901, and died March 27, 1977. They had moved north from Hayward, California, to raise their five grandchildren after the storm. It was a familial task they never bargained for in what should have been their retirement years. It was a responsibility they took on willingly, however, after prevailing in a legal custody battle with the maternal grandparents over who should raise the kids.

I lingered for a few minutes at the gravesite in a silence broken only by a dog barking in the distance and a faraway rumbling train. A few tree branches littered the cemetery grounds, reminders of a wind and rainstorm that swept through Western Washington two days before. We were entering the stormy season, which was ushered in so violently and lethally by the Columbus Day Storm.

12
Lions in the Wind

As the Columbus Day Storm raced through the south Puget Sound region around 8 p.m., winds gusted at 85 to 90 miles per hour in Spanaway, Washington, a modest suburban neighborhood just south of Tacoma. The beastly winds drew many of the uninitiated outside to experience the wind's energy and witness the damage all around—falling trees, shattered windows, downed power lines and fences, garbage cans tumbling down the street, and street signs and chunks of roof slicing through the air. Granted, stepping outside to be exposed to a major league weather disaster is not advised. But it's easy to see why people did it, given the rarity of the moment. The people of the Pacific Northwest had not seen a windstorm of this size and strength in recorded history. The storm strengthened as Charley Brammer, age seven, joined his parents, Ray and Mary Brammer, outside their Spanaway home. The parents wanted to check the condition of their roof, and Charley wanted to play in the wind. Their lives were to be changed forever.

Flashlight in hand, Ray Brammer shined a beam of light skyward to survey the just-replaced composite shingle on the roof. The sight was a bit unsettling: some of the shingles were standing on end. But the roof was holding together despite the windy assault. Charley, a spirited young boy just falling in love with baseball and soccer, was preoccupied by the wind in a more fanciful way. He spread his arms wide, scrawny wings on a flesh-and-blood airplane body, and waited to see if the gusts would lift him into the air. What young boy or girl doesn't at some time dream of flying? These are really strong winds, Charley thought, strong enough to make the dream of flying come true.

In a nearby field, a large animal paced in circles, agitated, confused about how to handle its newfound freedom. Charley's mother saw the outline of the animal in the darkness. She thought nothing of it, assuming it was a big dog. It was about time for the Brammers to return to the safety of their home. Charley's wind-aided flight turned into a stumble. The mysterious animal sensed his vulnerability and ran at him, eating up the distance like an animal on the hunt. The beast reached Charley and knocked him to the ground. "At first I thought it was our dog, Whitey—a Samoyed," Brammer recalled more than fifty years later during an interview at the Cabela's store in Lacey, Washington. "The next thing I knew I was being bit around my eyes and face." Charley let out a high-pitched scream and his parents came running. Blood poured into the young boy's eyes, which prevented him from seeing his attacker, then or later.

At first, Ray Brammer didn't know what the animal was either. He smacked it with his flashlight and his wife pulled off a slipper to beat the animal, too. The clubbings did nothing to stop the attack. Brammer grabbed the creature around the neck and tried to pull it off his boy. Then he saw the feline face of an African lion. "My God," he thought to himself. He realized he was in a fight for his young boy's life. A surge of adrenaline coursed through the father's veins, and he managed to pry the lion off his son. "Run for the house," he cried to Charley. Charley ran away from the house, away from safety. "I couldn't see, my eyes were so bloody," Brammer recalled.

His parents screamed out directions in a desperate bid to guide him back toward the house. By then, the two-hundred-pound lioness was free from Ray Brammer's grasp. Her long, loping strides ate up the ground between her and her prey. She pounced again and knocked Charley onto the concrete porch steps, opening an ugly gash on his forehead. Charley's parents were in frantic pursuit. Again, the father had to wrest the lioness away from her prey—his son—long enough for the bleeding, blinded little boy to scramble inside. The lion and father disengaged, and the parents dashed into the house. Once inside, they knew they had to get their injured son to the hospital. But outside they could see the lion pacing by the car, searching for wind-whipped scents of Charley. "The lion had the taste of Charley's blood in her mouth," Ray Brammer said as he reflected on that windy, frightful night.

Inside the sanctum of the home, Ray wrapped Charley's head in a towel to wipe away the blood, then pulled the towel away for a closer look, paying close attention to his son's eyes. Charley's mother took one look at her son's swollen and bloody face and started to scream, adding an anguished dimension to the winds that shrieked outside. Charley sat there, more in shock than in pain. Ray Brammer picked up his bloodied son and headed out to the car. But he beat a quick retreat when he saw the lion circling the vehicle. Trapped inside his own home, he called a neighboring friend, Frank Farley, for help.

Brammer told his friend that his son had been attacked by a lion, and that he needed help getting his wounded boy to the hospital. Farley couldn't believe what he had just heard. He thought it was a joke. He wondered if Brammer had been drinking. The father said he was dead serious, not drunk. He pleaded with his neighbor to come to their aid right away. Brammer's anxious tone was enough to convince Farley that this was no joke. Acting on an impulse, Farley grabbed a baseball bat and made the quick drive over, parking behind the house near the back door. Charley remembers his father snatching him up and hustling him to the car. But as his father stuffed him in the front seat, the lion reappeared from the front of the home, determined to pull his wounded prey from the vehicle. The lion sprang, and the father was locked in another life-and-death wrestling match with a lioness that wanted to maul his son. Ray Brammer fought his way into the car between his wounded son and the lion. He used the handle of the bat and his feet to push the lion away, then slammed the car door shut. Outside in the howling wind, the lion was once and for all separated from his prey.

The car sped away, but the drive to Tacoma General Hospital in Tacoma was a harrowing one. The street and traffic lights were darkened by widespread power outages. Downed trees and tangled power lines turned the route into a dangerous, electrified obstacle course. Thirty minutes later, the car pulled into the hospital emergency room parking lot. Inside the hospital, a nurse cleaned the face wounds and a doctor went to work stitching up the damage. Charley's left eyelid hung by a thread of flesh. A long gash within a finger's width of his right eye was another brush with blindness. The deep cut on his forehead required stitches, too. The doctor remarked later that Charley was lucky to still have his eyesight.

As Charley prepared to head home that night, he encountered Tacoma *News Tribune* photographer Wayne Zimmerman near the hospital entrance. Zimmerman was out chronicling storm damage all over town and stopped at the hospital in search of storm victims. A dazed young boy who resembled a battered prizefighter stood in front of him. His left

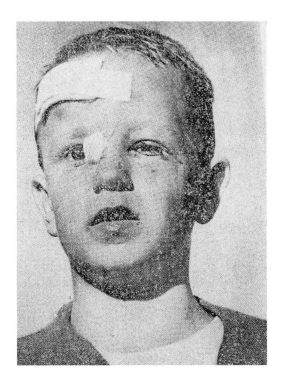

Charley Brammer, age seven, was mauled by a lion during the Columbus Day Storm, but escaped without life-threatening injuries. Photo by *Tacoma News Tribune* photographer Wayne Zimmerman.

eye was swollen and stitched and his right eye was partially obscured by a bandage abutting his swollen nose. A bruised and protruding right forehead was heavily swathed in gauze and tape to cover the stitches there. Zimmerman snapped a quick photo. "He told me my picture was going to be on the front page of the newspaper the next morning—I was on cloud nine," Charley Brammer remembered. "But the photo ended up on page 2." Speaking to a TNT reporter the next day, Mary Brammer marveled at how composed Charley remained through the whole ordeal. "He was the calmest of all of us," she said at the time. "He just said he wished lions had never been invented."

Where did the lion come from? Was it an escapee from some nearby zoo? Part of a traveling show of wild animals set free by the hellacious storm? No and no. The lionesses belonged to Joseph and Margaret McCallister and their son, Uwe, who lived a few blocks down the street from the Brammers. Charley remembered seeing Uwe McCallister walking the lioness twins down the street that past summer, secured by two precarious leashes. From the safety of the front porch, it was a sight to behold for the young boy. The singular parade was something exciting and out of the ordinary, but disturbing, too. The McCallisters kept their exotic pets in their backyard, in a twelve-foot-high wooden enclosure—one of the first things the fierce winds demolished as they swept through the neighborhood.

Where did the McCallisters get the lions? What would seem to be an obvious question has an elusive answer. Newspaper accounts and lawsuits after the mauling are silent. I had no luck locating members of the McCallister family for an answer. Ray Brammer remembers hearing someone say they acquired the lions from the Tacoma-based Point Defiance Zoo & Aquarium. Zoo officials said they found no record of any lion transaction between the zoo and the McCallisters. "In fact, we have no records indicating any transfer of lions to private parties," said Kris Sherman, public relations coordinator for the zoo. The zoo exhibited lions from 1915 to 1980. The numbers ranged from two lions in 1915 to ten cubs born in captivity in 1967, according to incomplete records supplied by zoo officials.

Perhaps it's just coincidence, but the Tacoma *News Tribune* carried a story in its August 21, 1960, edition, featuring twin lionesses born on July 23 and about to make their public debut. Zimmerman captured a shot of the twin sisters with their cocked heads, adorable inquisitive eyes, and oversized paws. The twin lionesses that escaped the McCallisters' crude enclosure the night of the storm were described as immature lionesses named Tammy and Sissy. Tammy had been declawed and defanged. The one that attacked Charley—Sissy—still had all of her predatory tools. Shortly before she mauled the young boy, she had pounced on another neighbor, Helen Sullivan, who nearly lost an ear in the attack.

While the Brammers were at the hospital, Pierce County sheriff's deputies were dispatched to the neighborhood. They encountered Tammy near the Brammer residence. Meanwhile, Sissy had returned to her enclosure and was captured by Uwe McCallister, who brought her to the sheriff's deputies to be put down. "They shot both lions in the alley about two blocks from our home," Ray Brammer remembered. A neighbor retrieved one of the shotgun shells used to slay the lions and gave it to Charley. "I had it for a long time," Brammer said.

Two months after the mauling, Ray Brammer filed a civil lawsuit in Pierce County Superior Court against the McCallister family, seeking $10,693 to compensate for the physical and emotional scars his son would carry with him for life. Initially the McCallisters rejected the claim, calling the unprecedented, severe windstorm an act of God that they could not have foreseen or guarded against. The court documents shed no light on why the McCallisters had two lions in their possession. This much is clear: there was no state law in place at the time to forbid the private ownership of exotic, dangerous pets. In fact, Washington State lawmakers didn't pass a law banning the private possession, breeding, and contact with all big cats, including lions, until April 2007, nearly forty-five years after Charley Brammer and Helen Sullivan were mauled.

The law in part was an outgrowth of a scathing investigation by the Animal Protection Institute in the summer and fall of 2005. The nonprofit institute, based in Sacramento, California, formed in 1968 to respond to animal cruelty and exploitation. In 2007 it merged with Born Free Foundation and became Born Free USA. During the 2005 probe, investigators visited private homes, federally licensed roadside zoos, and menageries that were home to exotic animals in North Carolina, Ohio, and Washington—three states at the time lacking laws to address private ownership. The probe uncovered numerous cases of children placed at risk of attacks, inadequate shelters for the animals, rough treatment, and overbreeding. The sting of publicity prompted the Washington State legislature to finally take action.

In January 1963, the McCallister family's infatuation with African lions caused them more legal woes. Helen Sullivan filed her own claim for damages, which reads like this:

> The plaintiff was painfully and brutally injured, requiring hospitalization at Madigan Army Hospital, and her injuries consisted of numerous lacerations of the scalp, neck and back, which required suturing, and that as a result of said injuries [the] plaintiff suffered severe pain and suffering and severe shock and fright and that she suffered organic brain trauma and an emotional reaction which will be permanent in nature, and she has suffered a permanent disability and permanent impairment of her earning capacity.

Sullivan's attorney argued for a settlement of $55,000. Again, the McCallisters leaned on their "act of God" defense. On February 14, 1963, the two lawsuits were consolidated into one. A jury trial in Pierce County Superior Court was set for June 5, 1963. Pretrial maneuvering by the plaintiffs and defendants continued over the next three months. On May 13, the defendants asked for and received a motion to delay the trial. On May 20, the Brammer's attorney, Thomas C. Lowry, entered into the court record an assessment by Dr. E. E. Benfield, a Tacoma-based plastic surgeon, of Charley's facial scars on the forehead and right side of the nose, suggesting they would need repair, but that the surgery should be delayed until Charley reached "full facial growth" around the age of twelve.

At some point during the legal fight for damages, Uwe McCallister came over to the Brammer home and talked to Ray Brammer. Brammer described him as a teenager, and, according to court records, his father was in the army, stationed in Germany. "He came over and said, 'We don't have any insurance and we don't have much money,'" Ray Brammer recalled. "I kind of felt sorry for him." The Brammers talked it over. They decided to forgo trial and accept a settlement of $3,250. After attorney fees, the Brammers pocketed $2,085. Most of it was deposited in a savings account for Charley to tap into when he turned eighteen. The money doubled with interest over the next ten years. It was enough cash for the eighteen-year-old to buy a 1964 Ford Galaxy. The case involving Sullivan and the McCallisters never went to trial either. Court records go silent after the Brammer settlement, but Sullivan must have settled with the McCallisters, too.

In the years that followed, Brammer doesn't remember nightmares about the lion attack, or holding a permanent grudge against lions. He thinks his mind is clear because he never saw the beast that attacked him. His dad isn't so sure that his son escaped without some lingering effects from the mauling. For years, he didn't want to be around big dogs, his father said. "I don't think about it—it doesn't bother me a bit," Brammer, now a middle-aged man, told me as he munched on a hamburger and french fries in the cafeteria of the Cabela's sporting goods store in Lacey, just a few feet removed from a stuffed male African lion. "I used to tell the story to people when they'd ask me about my scars." Nowadays, he rarely talks about it, and it's been that way for years. "We were married a year before I ever heard the story," said his wife, Debbie.

The deferred plastic surgery referenced in the court papers never happened. The scars were visible as the young boy matured into a teenager and adult. But the only scar that really bothered him was the one on his forehead, inflicted when he crashed into the front porch steps. Because of nerve damage, any contact with a foreign object—think baseball cap—triggered pain akin to an electrical shock. But the shock has subsided to a tingle, and the scars about the eyes and nose of this tall, lanky man are obscured by the wrinkles that arrived with time. Add a weathered look to an unassuming guy with the same sandy hair he had the night of the attack, and one can link the two images, one shortly after the attack, and one fifty-plus years later.

Most comfortable in work boots, jeans, and suspenders, he likes to hunt deer and elk and play golf. He earns his living as a plumber at the Western Washington Fairgrounds in Puyallup, Washington, a sprawling complex that is home to the state's biggest fair. It comes as no surprise that his dad was a plumber, too. They embrace their unique bond brought about by a freakish storm and a freakish encounter with a lioness. It's hard to imagine one much stronger. "We've always been close," acknowledged the elder Brammer. "But I just did what any father would do."

Not far from the Point Defiance Zoo & Aquarium, Terry Terrien, fourteen, boarded the SS *Skansonia*, the 7:30 p.m. ferry bound from Tacoma, Washington, to Vashon Island Friday night, October 12. It was about a twenty-minute ride from the mainland to the island, one of many that dot the fjord-like, fingered inlets of Puget Sound. The young teenager was headed with a family from his church to Camp Burton, an island church retreat. The car rolled onto the ferry, an oval-shaped, 1929 wooden vessel with a superstructure built to hold thirty-two cars and some 465 passengers. Terrien noticed that the gusty winds buffeting the dock were increasing in intensity, creating a deafening howl. The ferry's departure coincided with the arrival of the Columbus Day Storm to Tacoma, a steep-sloped city that plunges into Commencement Bay just beyond the mouth of the Puyallup River. Dubbed the City of Destiny in the 1880s by its railway-terminus-seeking, optimistic founders, its true destiny was to sit in the economic and cultural shadow cast by Seattle some thirty miles to the north.

The shipmaster's whistle blast announced the ferry was ready to embark on what would be a harrowing ride that Terrien would never forget. Crossing the bay had never been tougher. The winds reached gale force, creating ocean-like swells and frothy waves that muscled up against and crashed over the stern of the 165-foot-long ferry. Nine minutes after leaving the Port Defiance dock, a crewmember passed through the ship's observation lounge and deck, telling passengers the ferry was about to arrive at the island landing, named Tahlequah, an unincorporated community at the south end of Vashon Island, a long way from the capital of the Cherokee Nation in Oklahoma that laid claim to the name first.

Terrien opened the door at the foot of the stairway that descended to the car deck. He looked up to see what appeared to be a tall, pyramid-shaped object, cast in the light of the ferry's floodlights. It was the piling guide to the dock, and the ferry was approaching it much too quickly. "The captain slammed the ferry into reverse," Terrien recalled five years later, as a journalism major at Linfield College in McMinnville, Oregon. "His efforts were futile. We hit with such force I was nearly thrown to the deck. I thought [for] sure it [piling guide] would topple, but it sustained the blow."

The ferry bounced off the dock approach. The skipper of the *Skansonia* regrouped and tried another run at the moorage, but missed to the left, and was headed for a crash-landing on the beach. A crewmember on the bow of the car deck waved his arms wildly at the wheelhouse, trying to capture the captain's attention. He shouted, too, but his words of warning were swept away in the wind. The captain threw the diesel engines into reverse just in time to keep the ship from running aground. Once again, the ship slowly reversed back into the roiled waters of the bay, fighting the headstrong winds as waves broke over the side of the ship and poured over the deck.

On the third attempt to follow the landing lights to the ferry dock, the ship missed to the right. The ship's captain maneuvered the ferry back out into the bay for a fourth try at safe harbor. This time the ferry was aligned with the dock lights. Just then, the dock lights blinked out, a victim of an island-wide power outage. With no lights to guide him, the captain turned the ship around into the teeth of the wind in a desperate, uncertain bid to return to Point Defiance.

Terrien climbed back up the stairs to sit out the return voyage in the ferry's lounge with twenty other passengers. The waves struck the bow and starboard side of the ship, pitching the ferry violently from side to side. Leaving the partial windbreak provided by the island, the ship rocked over on its port side, then slammed back down into the water with a thundering blow. Terrien cast a glance around the lounge. A mother and her child donned life jackets, preparing for the worst. A small cluster of business commuters in suit and ties cracked nervous, half-hearted jokes. Most passengers, including Terrien, were alone with their dark thoughts as the wooden vessel creaked and groaned, jarred over and over by the violent pitch of the waves. "The feeling of helplessness was the greatest suffering of all—to just sit and be thrown about, not able to lift a finger in my own aid," Terrien recalled. "I felt as if I were a distress note within a forgotten bottle being tossed to and fro on the ocean seas waiting and praying for someone to pluck me out."

The ship's engines strained to move the ferry along. A fire broke out in the ship's engine exhaust stack, a sign that the engines were being over-worked in the struggle to make headway against the wind. Crewmembers worked to extinguish the flames while the passengers cast blank, wide-eyed stares at the crashing waves and dark sky. Suddenly, Terrien heard a passenger say the lights of Point Defiance were coming into view. Passengers scampered to the windows and searched for a view of the ferry landing through the gusts of spray and wind. "I sank back down into my place, dizzy with relief," Terrien wrote five years later. "Passengers began to smile and talk freely for the first time. The room grew warm."

The ferry eased into the landing, ending a nearly hour-long trip across the stormy bay. Passengers hustled down to the car deck to find their once neatly parked vehicles in disarray from the storm's jostling. They drove off the ferry, relieved to be on land again, but anxious to see what storm damage awaited them at home.

13

It Happened at the Fair (Buon Gusto)

It took the storm less than thirty minutes to shoot north from Tacoma to Seattle, the region's largest city and proud host to the 1962 Century 21 Exposition, better known as the Seattle World's Fair. Duff Andrews, twenty-two and a dental school student at the University of Washington, hesitated a moment before stepping into the freight elevator serving the fair's iconic Space Needle. He and coworker Robert Harvey, nineteen, had been told by their supervisors to return to the five-hundred-foot level to roust from the bar four boozy patrons, who were ignoring orders to evacuate the Space Needle in the face of fierce Columbus Day Storm winds pummeling the fairgrounds.

Andrews figured he made twenty thousand trips during the fair on the three Space Needle elevators, one of which wasn't delivered and installed until the day before the fair opened on April 21, 1962. He was about to take his most memorable ride. "Before we got back in the elevator, I looked at the Space Needle and it was twisting in the wind," said Andrews, a retired dentist living in the San Juan Islands. Despite the disturbing condition of the fair's prime attraction, the two employees hopped in the freight elevator and started their ascent, a ride that typically took one minute.

The three Space Needle legs drew close together about halfway up before flaring back out to support the five-story disc that housed the restaurant and observation tower. It was near the midway point that the elevator malfunctioned and stopped. "We were very scared at first," Harvey told a newspaper reporter the next day. "The wind was really going." Andrews recalled the elevator bobbed up and down on its seven cables, each capable of supporting the elevator on its own. But the elevator lights were still working, and they used a telephone in the elevator car to report their plight.

The back panel on the elevator opened up to a stairwell in the central core of the needle, but a cyclone fence stood between the door and the stairs. Andrews and Harvey weren't going anywhere. While they waited to be rescued, Andrews pulled a deck of cards out of his pocket. "You never know when a deck of cards will come in handy," Andrews said. "We had lights, so we sat on the floor of the elevator car and played a game of gin rummy." The elevator car was bouncing up and down, one to two feet at a time, but the card game continued, with Harvey winning. Andrews said it was like bobbing in a boat, suspended in midair. The two coworkers just got used to it. A rescue crew climbed the stairs, cut an escape hatch in the fence, and helped the two young men out of the elevator for the long walk down the stairs. "We were glad to get out although we knew that nothing bad was going to happen to us," Harvey said the next day, his youthful bravado restored by a night's sleep.

Undaunted by ominous weather reports, Italian Americans and others of all ages flocked on October 12 to the 1962 Seattle World's Fair. It was Columbus Day at the fair in honor of Italian New World explorer Christopher Columbus. Nights, days, even full weeks of recognition were bestowed on states, nations, cities—even the Camp Fire Girls and the American Automotive Association—throughout the 184-day run of the fair. Before the night was over, Columbus would share his name and notoriety with one of the strongest nontropical windstorms to ever make landfall in the lower forty-eight states. The seventy-four-acre fairgrounds, including the Space Needle, stood in the path of the storm.

A banner headline on the front page of the Friday afternoon *Seattle Daily Times* foretold of a dangerous storm packing winds up to 75 miles per hour on the Washington and Oregon coasts. The wind forecast for Seattle, the country's northernmost big city, tucked between the jagged Olympic mountains to the west and the Cascade foothills to the east, nestled on the shores of Elliott Bay and Lake Union, with mighty Mount Rainier looming sixty miles to the southeast, was much less severe, but still high enough to warrant attention: 35 to 52 miles per hour as night fell Friday evening. The forecast proved to be much too conservative.

The steady stream of tourists, who had helped make the Century 21 Exposition a success and sent fair attendance beyond 9.6 million, had returned to their far-flung homes by early October. In just ten days President John F. Kennedy was scheduled to arrive for the fair grand finale, a noteworthy ending to a worldly celebration that put Seattle on the map and set the stage for an uncertain "space age," a future filled with hope and promise of technological advances, but also a future full of dread and fear triggered by an escalating Cold War between the Soviet Union and the United States, one that could lead to nuclear war.

On this early fall day on the cusp of a rainy season that extends into June in the Pacific Northwest, some forty-one thousand people, mostly from the Pacific Northwest, pushed through the fair turnstiles, reclaiming the fair as their backyard playground now that summer vacationers from all over the country and the world were back home, their children ensconced in school. Local residents were accustomed to a shot or two of stormy spring and fall weather, and the fair had already weathered an April 27 windstorm with winds reaching 70 miles per hour. They weren't about to let some windy weather deny their fun in the fair's final days.

The Space Needle, the Monorail, and the science and technology theme of the fair grew from a Cold War response to the Soviet Union's initial space-age superiority, including the launch of *Sputnik 1* in 1957, the first satellite to orbit the Earth. The Soviets had scored the first space-age punch, placing the United States in a catch-up mode. The Seattle World's Fair became an ideal venue for the US government and private companies to display their own high-tech advances and desires.

Above all other fair exhibits, pavilions, and displays, the Space Needle was the fair's "wow" factor, that extra boost that Seattle civic schemers and dreamers needed to sell the fair to other nations (you want to be a part of this!). The media and the deep-pocketed doubting Thomases in the Pacific Northwest at first had a hard time envisioning Seattle transformed from the blue-collar gateway to Alaska to the futuristic gateway to the stars. The doubters initially included the almighty Boeing Company, the homegrown airplane company that piloted the region's economy. But the Space Needle in all its spindly glory was destined to be the launchpad and the bridge to the future. Even Seattle's World Fair colors were true to the space-age theme: Astronaut White for the legs, Orbital Olive for the core, Reentry Red for the halo, and Galaxy Gold for the top.

Columbus Day festivities kicked off noon Friday in the shadow of the Space Needle at the Plaza of the States, a flag-filled daily gathering place for kudos and glad-handing, music and speeches. The plaza was a late addition to the fairgrounds at the urging of Washington governor Albert Rosellini. The first Italian American elected governor in a state west of the Mississippi River, Rosellini was ebullient, in his element, his cheesy jaw-jutting smile beaming at the crowd as he introduced the guests of honor, including Dr. Paola Rota, Italian consulate in Seattle, and Victor Rosellini, a Seattle restaurateur who served as vice-chairman of the state's World's Fair Commission from 1957 to 1963 and received a distinguished service award that Columbus Day from his cousin, the governor. A keen supporter when the fair was in its formative years, and frequent fair visitor, the governor's trademark smile had been on display numerous times on the fairgrounds, including in September when he accepted a Tennessee ham from rock-and-roll sensation Elvis Presley, who was on hand for the filming of a forgettable movie called *It Happened at the World's Fair*.

Rosellini, a progressive Democrat, was in the midst of his second term as governor, an office he ran for five times and won twice. On Tuesday the week of the storm, he had unveiled a bill for the 1963 state legislature requiring seat belts in all cars sold in the state, beginning in 1964. Whether in the area of prison reform or mental health care or the creation of today's world-renowned medical school at the University of Washington, Rosellini was a social reformer at his best. The son of an Italian immigrant, the former collegiate boxer had spent his early years as an attorney, at times defending the pimps, prostitutes, and racketeers that made up Seattle's underbelly in the 1940s, 1950s, and early 1960s. In a public and political life filled with contradictions, he also served as a King County deputy prosecutor and a crime-fighting state senator before ascending to the governor's mansion in Olympia in 1956. Rumors of his cronyism and underhanded dealings dogged him through seventy years of public life, but he was never charged with a crime. That didn't stop the FBI under Chief J. Edgar Hoover from compiling a 779-page investigative file on Rosellini, a file filled with allegations and gossip, and not made public until Rosellini's death at the age of 101 on October 10, 2011.

At the northeast corner of the fairgrounds was Show Street, the adult entertainment venue at the otherwise family-oriented fair. Frank Colacurcio, who would become one of Seattle's most notorious strip club owners and racketeers in the post–Seattle World's Fair era, got his start in the nightclub business on Show Street. He owned and managed the Gay Nineties–themed Diamond Horseshoe. In August he ran afoul of authorities for employing four underage girls as dancers at the club. He received a $500 fine and a six-month suspended sentence, one of Colacurcio's many run-ins with the law. Twenty years earlier, he had been convicted of a statutory rape charge, despite the efforts of his defense attorney, thirty-two-year-old future governor Albert Rosellini.

As night fell on the "Emerald City," the fairgrounds still bustled with twenty thousand patrons. Many called Seattle home, this city whose steep hillsides had been sluiced and shoved into the bay—fifty million cubic yards worth—fifty years earlier to create more waterfront and room for unfettered, unmitigated growth, a city barely a hundred years old; a city with economic boom-bust cycles akin to the changing seasons; a city just beginning to wake up to traffic gridlock, pollution, and loss of open space; a city named after a Native American, Chief Seattle, whose people were banned from living in the frontier city that became the largest on the continent (pop. 655,000) to bear a Native American's name; a city that gave birth to Boeing airplanes, Microsoft computers, Starbucks coffee, Red Hook beer, and Amazon online sales. Few fair patrons or anyone else in town knew that the storm of the century was fast approaching.

By 6 p.m., wind gusts strong enough to knock people off their feet reached the fair grounds. The Washington State Pavilion, later named the Seattle Center Coliseum, was closed at 7:30 p.m. by fair officials, who feared wind gusts might blow out windows there. The Plaza of the States was no longer a welcoming place: the wind played a discordant, frightening song on the halyards of the plaza's fifty-four flag poles.

The hundred-foot-high Gothic spires wobbled in the wind at the US Science Pavilion where five months earlier US astronaut John Glenn, flush with celebrity earned from his February orbit around the earth three times in a little under five hours, sat cross-legged on the carpet of the Boeing Spacearium (Boeing Company doubters having come to their world's fair senses), next to US Senator Warren G. Magnuson (D-Wash.), to watch *The House of Science*, a Science 101 film that must have seemed pedestrian to the nation's first space traveler. Always competing, the Russians sent Soviet cosmonaut Gherman Titov to visit the fair five days before Glenn arrived. In August of 1961, Titov was the first person to spend a full day in space and the second to orbit the earth—Soviet cosmonaut Yuri Gagarin was first.

All over the fairgrounds trees snapped, banners and signs flew through the air, and children clung to the legs of their parents, who ushered them to the fair exits into an uncertain night. Jack Gahagan, forty-three, a worker in Gayway, the fair's space-age-themed amusement center, fell eighteen feet to the ground from a platform while trying to secure the Fun-O-Rama display panels. He was admitted to Seattle General Hospital, but suffered only bruises. Nearby, 110 feet above the ground, Loren Rankin and his wife, Ruth, were trapped for two hours at the top of the Ferris wheel, an amusement ride that had debuted at another world's fair—the World's Columbian Exposition in Chicago in 1893.

The Chicago fair was approved by Congress in 1890 to celebrate the four-hundred-year anniversary of Christopher Columbus's arrival in the New World in 1492. Construction delays caused the Chicago fair to open a year behind schedule, but that's not out of the ordinary: the Seattle World's Fair was originally set to open in 1959 to celebrate the fifty-year anniversary of the Alaska–Yukon–Pacific Exposition held in 1909 on the University of Washington campus, which consisted, at the turn of the twentieth century, of three buildings tucked in an old-growth forest near the shores of Lake Washington. That wildly successful fair was two years late in celebrating the ten-year anniversary of the 1897 Klondike Gold Rush. Fair-opening dates are moving targets, elusive at best.

The novelty of being stuck at the top of a Ferris wheel in howling winds wore off quickly for the Rankins. Their seat swung three feet to a side, buffeted by the wind. "The lights were out on the Ferris wheel and the fairgrounds and all over the city," Ruth Rankin recalled, although other accounts of the night suggest the lights stayed on at the fair. "It became rather frightening."

Among the nighttime fair attendees were Seattle KIRO radio disc jockey Jim French and his wife, Pat. They attended a reception at the top of the Space Needle for members of the Seattle media and their spouses. The faithful chroniclers of the fair's successful six-month run were in a festive mood. Word began circulating that a big storm had hit Portland, knocking down the television transmission towers there, French recalled. "It didn't sound real," French said. "Nobody believed it at first because we don't have weather like that in the Pacific Northwest."

Maybe French had heard UW atmospheric sciences professor Phil Church utter these reassuring weather words on the eve of the Seattle World's Fair, words that fit nicely in a cadence borrowed from the Lord's Prayer:

We residents of Western Washington
Are thankful to be cheated
From witnessing or being subjected to
Nearly all the vicious display
Of huge atmospheric energy bursts.
Such as hurricanes, tornadoes, severe thunderstorms, dust storms,
      cloud bursts, blizzards, ice storms cold and hot waves
that people in other sections of the country must endure.
Amen.

Meanwhile, some 174 miles south of Seattle, the radio team at KGW 620 in Portland, including meteorologist Jack Capell, was fielding calls from frantic listeners asking how long the winds would last, and was it true another storm was approaching from the south? "I was trying to keep them informed on whether the worst of the storm had passed," Capell said years later. "I guess you could say I was the calming voice."

The winds continued to stiffen in Seattle, and the Space Needle began to sway ever so slightly. That's what it is designed to do in a windstorm, move about one inch for every 10 miles per hour of wind velocity. The Space Needle was built to withstand winds of 200 miles per hour, more than twice the speed of the winds that raked it that Friday night, and double the building code requirement. The needle was also designed to escape serious structural damage in a magnitude 9.1 earthquake, which suggests it would likely remain erect in a much-feared Cascadia subduction zone earthquake, which features a tectonic battle—the oceanic plate diving under the continental plate off the West Coast, locking up, then breaking loose. These cataclysmic earthquakes rip the Pacific Northwest apart every few hundred years. Seismologists and geologists like to point out it's been more than four hundred years since the last subduction zone earthquake.

French wasn't concerned by the swaying Space Needle. His confidence in the structure had grown over the many lunch breaks he spent near the construction site, watching the enduring symbol of the Seattle World's Fair emerge from a concrete foundation thirty feet below ground surface and 120 feet square. It took nearly five hundred cement trucks twelve hours to fill the hole in what was then the largest continuous pour of concrete ever attempted in the West. In just four hundred days, the Space Needle evolved from an architectural and engineering plan to a completed project, looming 605 feet over the city skyline. The marketing team at US Steel dubbed the Space Needle "The 400-Day Wonder."

The Space Needle grew from a doodle on a placemat or napkin by Western Hotels executive and world's fair mastermind Eddie Carlson after a visit to a restaurant atop a 711-foot-high broadcast tower in Stuttgart, Germany, in April 1959. With its tripod legs and saucer-shaped top, the Space Needle was the tallest building west of the Mississippi when the fair gates opened on April 21. The needle's two public elevators—a third service elevator was set aside for freight and dignitaries—whisked nearly twenty-thousand people a day from ground level to the third-highest observation tower in the nation in a mere forty-three seconds, at a speed of 10 miles per hour, the speed of a raindrop falling. People stood in line for hours to pay one dollar for a ride to the top.

The night of the windstorm, elevator service to the needle's observation deck and Eye of the Needle Restaurant ground to a halt about 7:30 p.m., but diners were allowed to finish their meals. The cautionary closure of the Space Needle was a first. Loudspeakers scattered around the fairgrounds blared out a somber message: Winds topping 80 miles per hour are expected soon. Leave the premises and seek shelter immediately, the message sounded. Some fairgoers heading for the exits used their umbrellas as shields to guard not against rain but, rather, flying debris and glass. French and his wife walked briskly across the fairgrounds to their car. A neon sign sailed through the air and crashed at their feet. Their pace quickened. The wind was tremendous and frightening. French took note of a newfound truth: the Pacific Northwest is not immune to hurricane-force windstorms.

French and his wife walked through downtown Seattle past the Monorail, a 1.3-mile-long elevated train that linked downtown to the fairgrounds. It was surpassed only by the Space Needle as the most iconic symbol of the space-age fair. The sound of a four-engine plane drowned out the howling wind. The plane was flying south, well below the Federal Aviation Administration's five-hundred-foot ceiling. French, a pilot, watched awestruck as the plane struggled to make headway into the onrushing southerly wind, traveling no faster than a car in noon-hour traffic through Seattle's city streets. Blue flames shot from the plane's engines. The pilot was probably on a desperate mission to land at Boeing Field south of downtown Seattle.

These were heady times at the Boeing Company. In just a few weeks, the aerospace giant would roll the first Boeing 727 off its assembly line in Renton, Washington. It was the first trijet (an aircraft with three engines) introduced to commercial airline service, and that plane became an aerial workhorse, the top-selling jetliner for the first three decades of the jet transportation era. French described the harrowing ride east across Lake Washington to reach their Bellevue home. Wind-whipped waves crashed against the lake's floating bridge. "It was like surfing in a Chevrolet," he said.

Back at the Space Needle, not everyone chose to linger over their $7.50 salmon or steak dinner served with a drink. Not everyone found comfort sitting on a dining floor surface that rotated full circle every hour. Some left the needle under duress as the Space Needle swayed and twisted like a ship in stormy seas. The rushing wind turned the needle's sturdy legs into giant tuning forks, whipping up an eerie, whistling sound. A few hungry Space Needle refugees happened upon Teepee Salmon Barbeque, not far from the big gate entrance to the fairgrounds at Fifth Avenue and Mercer Street. Ralph Munro, a student at Western Washington College in Bellingham, cooked there on weekends after spending a summer working at the fair. He was the grandson of Alexander McKenzie Munro, a Scottish stonecutter who helped rebuild Seattle after the Great Seattle Fire of 1889 and, thirty years later, shaped sandstone to construct the state capitol building in Olympia. The younger Munro went on to a noted career in politics, a Teddy Roosevelt–style Republican and a chief aide to Dan Evans, who beat Rosellini in the 1964 governor's race. Munro, a tall, broad-shouldered guy with a squarish face and folksy sense of humor, served as Washington's secretary of state from 1980 to 2000.

One of Munro's harried customers told him she saw a pastry case fly open in the Eye of the Needle Restaurant, spilling pies on the floor. "Folks were really upset," Munro said. Many of them had to climb down the stairs when the elevators were shut down as a safety precaution. That's 848 steps to ground level. The barbeque closed at 8:30 p.m. With the trolley buses idled by the storm, Munro started walking through the streets of downtown Seattle. A big display window at a downtown department store popped out and exploded against a parking meter as he was about to pass. He was headed to the Pier 52 ferry terminal to catch a ride across Elliot Bay to his family home on Bainbridge Island. This busiest ferry terminal in the nation was built by Scottish engineer James Colman in 1882. The wooden structure succumbed to the flames that destroyed most of the city in the 1889 fire, the same year the Washington Territory achieved statehood. The dock was quickly rebuilt.

At 9:15 p.m., Washington State officials shut down the state-run ferry system serving Puget Sound and the San Juan Islands. The ferries could not navigate the frothy, white-capped bay. Munro joined two hundred or so stranded commuters sleeping on wooden benches in the domed ferry terminal waiting room, which had two inches of water on the floor from a minor storm surge during the evening high tide. The maritime storm scene in Seattle, a city surrounded by water, was summed up Friday night by a dispatch from the Coast Guard Search and Rescue Center there: "The winds are gusting up to 75 knots. . . . There are log booms and boats adrift all over Lake Washington. There are houseboats adrift in Lake Union and pleasure boats at Shilshole Bay. The water's just about too rough on Elliott Bay to get a boat out. We almost lost one of ours."

The dispatcher was referring to a sixty-four-foot-long Coast Guard tug that started rolling heavily after pulling out into the bay from Pier 39. The tug was forced to turn back. By 10 p.m. the winds had begun to subside, and Coast Guardsmen were back on the water, with orders to "round up everything in sight."

The search and rescue efforts were too late to save the recently retired *Chief Kwina*, a ferry that had worked the run from Bellingham in north Puget Sound to nearby Lummi Island. The empty ship sank at its mooring at Lummi Island in the early morning hours Saturday, after waves had pounded over the ship's superstructure for more than three hours. At the peak of the storm late Friday night in Bellingham, where winds neared 100 miles per hour, mariners reported waves more than thirty feet high in Bellingham Bay outside the city's Squalicum Harbor. One of those waves cleared the mast of a Coast Guard cutter that was responding to a distress call near Point Frances. Unable to make headway into the winds, the Coast Guard vessel had to retreat and tie up at a Port of Bellingham dock outside the impassable harbor breakwater until the storm subsided.

Ron Newell, a nineteen-year-old Bellingham native, watched the storm unfold from his grandparents' home at the foot of Sehome Hill, overlooking the Port of Bellingham, Squalicum Harbor, and Bellingham

Bay. "The waves on the bay looked like ocean swells," said Newell, who still lived on the shores of Bellingham Bay when we talked in 2015. Newell, also a student at Western Washington College in 1962, had more than a passing interest in the storm. On Columbus Day, he was a US Weather Bureau storm warning displayman for Bellingham Bay, raising weather warning flags in the day and lanterns at night that were attached to a 150-foot-tall steel tower next to his grandparents' home. The flags and lanterns alerted boaters to weather conditions on the water in an era when few smaller vessels were equipped with weather band radios.

Newell's grandmother, Bessie Newell Otto, had had the job since 1927, but she was aging and needed someone younger to hoist the flags and climb the tower for the occasional replacement of a light. On Friday morning, Newell had received a telegram from the Portland weather bureau office, directing him to raise the most severe whole gale flags, a warning of winds likely to reach 55 to 63 miles per hour, which turned out to be an understated forecast. Newell figured he was probably the first person in Bellingham to hear about the approaching storm.

The last time those two red flags with black centers had flown was in advance of the October 21, 1934, windstorm, which brought gusts of 60 to 70 miles per hour to Western Washington. As darkness fell on the bay, and the storm had yet to arrive, Newell didn't waver from the forecast. He raised a white light lantern bracketed by two red light lanterns, which was the nighttime warning of whole gale winds. It's safe to say at least some recreational and commercial boaters in Bellingham heeded the warnings and decided not to set sail that day.

The next morning Munro caught a 5 a.m. ferry to Winslow, only to learn that his mother, a schoolteacher, had suffered a heart attack in her classroom Friday before the storm struck. She died Saturday at Harrison Hospital in Bremerton. Munro said she had a preexisting heart condition and should not be considered a victim of the storm.

Just as state officials were shutting down the ferry system, most of the Seattle World's Fair experienced an emergency closure, too. The Food Circus, which was the old Seattle Armory, converted into fast-food stalls of cuisine from around the world, remained open. It served that night as an emergency shelter for fair visitors from communities south of Seattle hit hard by the storm. The other exception to the early fair closure was

Show Street, which blared on with its burlesque shows, vaudeville acts, and nightclub partying as if it was just another Friday night in the Emerald City. Around 10 p.m., fair security forces asked the Show Street clubs to shut down. Even then there were people in line outside, waiting to buy tickets to adult shows, oblivious to the weather extravaganza ripping through the city.

The windstorm marched north through the night, leaving fairground crews enough time to clean up the storm debris and reopen to a Saturday crowd of more than seventy-five thousand. The fair finale on October 21 didn't go off as planned. Two days before the fair ended, President Kennedy sent his regrets. He was suffering from a severe cold and under doctor's orders not to make the cross-country trip to Seattle, according to his press secretary, Pierre Salinger. Also cancelled was a Sunday breakfast at Seattle's grand Olympic Hotel with President Kennedy and two thousand ticketholders. It was a fitting venue: the 1924 Italian Renaissance hotel was the incubator site for the world's fair, dating back to the World's Fair Commission's first meeting there in 1955, chaired by Carlson. Planning breakfasts continued there, 7 a.m. daily, in the years leading up to the fair. The powerful and persuasive Magnuson tried to convince LBJ to fill in for the president, but, coincidentally, the vice-president had a debilitating cold, too.

What Kennedy and Johnson had wasn't a cold but rather a Cold War confrontation of epic proportions—the Cuban Missile Crisis. For several days leading up to the president's no-show in Seattle, he was huddled with his closest advisers and top military brass, trying to decide on a course of action to counter the Soviet Union's bold, taunting decision to place Soviet nuclear missile sites and bombers in Cuba, just ninety miles from US soil. While the fair played out, Kennedy pondered his options: invade Cuba, launch an air strike to destroy the missiles, or something slightly less aggressive. Much to the chagrin of the joint chiefs of staff—they wanted to strike Cuba by air with nuclear bombers and missiles—Kennedy chose a naval quarantine of Cuba—an act just short of a blockade, which would have been considered an act of war.

Kennedy went on television the day after the fair closed to address the nation, to tell us enough to scare us about how close we were to a nuclear exchange with the Soviets. The nation's Strategic Air Command, which controlled much of the nation's nuclear arsenal, was on DEFCON 2, the highest state of military readiness short of war. All those drop-and-cover exercises in school suddenly took on new meaning. Initially, Soviet premier Nikita Khrushchev swore he wouldn't back down. But Soviet ships honored the quarantine, and the Soviets agreed to dismantle their Cuban arsenal. In return, the United States vowed not to invade Cuba, a growing Communist stronghold under Fidel Castro's grip. In addition, the United States promised to remove its nuclear missiles from Turkey, which were within easy striking distance of Russia. Nuclear annihilation was averted.

I returned to the Space Needle observation tower in early May of 2014, a full forty-five years since my last visit. Like many Puget Sound residents, I take the enduring structure for granted. I'm used to seeing it in any visual sweep of the Seattle skyline, any advertisement selling Seattle merchandise, any television show before or during a Seattle Seahawks football game. When I visited the Space Needle it was no longer the tallest building on the Seattle skyline; five downtown skyscrapers towered over it, including the Columbia Center at 937 feet tall. But it is omnipresent, forever unique, and firmly entrenched in the twenty-first century, the one it was built to foretell.

The Space Needle hosts about one million visitors a year, making it the number-one tourist attraction in the Pacific Northwest. On this Saturday afternoon, the observation deck was crowded, but not suffocating. Asians dominated the scene, most of them young and most of them using their iPhones to shoot selfies. I forked over twelve dollars for a twenty-ounce Pike IPA, one of the many craft breweries that sprang up in town after Gordon Bowker and Paul Shipman launched Red Hook in a converted machine shop in Seattle's Ballard District in the early 1980s. I sipped my beer and soaked in the 360-degree sights. A cruise ship the size of a skyscraper flipped on its side steamed out of Elliot Bay. Draped in dreamy clouds, the Olympic Mountains folded and tumbled to the west into the Pacific Ocean, the breeding ground of enough foul weather to belie its name.

Downtown Seattle stretched to the south, a place where high-tech moneymakers are driving apartment rent and condo prices through the roof. The gentrified cityscape gave way to the industrial, polluted Duwamish River, a river named after Chief Seattle's tribe. There a tangle of industry is heaped on a contoured pile of dredge spoils that is the 395-acre Harbor Island, the largest man-made island in the country. Water, mountains, and blankets of conifer trees were everywhere, dissected and isolated by freeways, bridges, and tendrils of suburbia. The Bellevue sky-line on the east side of Lake Washington looked like Seattle's little brother, still growing into its steel, glass, and concrete high-rise body.

If there was a breeze blowing, I didn't feel it. If there was a fear of heights present in the animated, multicultural, observation deck crowd, I didn't sense it. Look at Mount Rainer, that crumbly old volcano whose summit I stood on thirty-five years ago and whose glaciers are receding in the time it takes to say climate change. If the Space Needle suddenly started to twist and sway in the clutch of hurricane-force winds, all these spectacular views would be soon forgotten. There would be a stampede for the elevators and the stairs, and I'm pretty sure I'd be swept up in the thick of it.

14

Terror in Stanley Park

The storm was enduring, interminable, the high-pressure wave still trying to fill the low-pressure trough, a game of atmospheric catch-up, which, when it happened, would extinguish the storm's might. But that meteorological equilibrium was still elusive, nowhere in sight. The winds pushed north another two hours and ninety miles from Seattle to Bellingham, Washington, reaching 98 miles per hour before a final assault across the Canadian border, an act that imprinted the storm on the landscapes and cityscapes of two nations.

Crossing the border into the Canadian province of British Columbia, the wind had a swift, unimpeded journey across the Fraser Valley and its broad, flat estuary. The Vancouver metropolis was just minutes away. As the clock struck midnight, the aging, raging storm reached the third-largest urban oasis in North America, Vancouver's Stanley Park, a 1,001-acre peninsula of thick, lush green forests surrounded by rocky beaches and marine waters, shaped similar to the Hawaiian island of O'ahu, if it weren't for a slender, terrestrial connection to downtown. The park's shoreline features a thirteen-mile seawall promenade with spectacular vistas of English Bay and Georgia Strait to the west and, northwest, Burrard Inlet to the east and Coal Harbour to the south. The park interior is home to majestic Douglas-fir, western hemlock, and western redcedar trees, some more than two hundred feet tall, many predating creation of the park in 1888. The park was named after Lord Frederick Stanley of Preston, then governor-general of Canada and an avid hockey fan, who donated the decorative punch bowl that is awarded each year to the champions of what would become the National Hockey League.

The park's popularity is the stuff of legend. When Vancouver incorporated in 1886, the city council's first order of business was to ask the government of the Dominion of Canada to bequeath to the city this treasured piece of ground that was at the time a military reservation. The mission was accomplished, sealing a marriage between the citizenry and the land that embodies a passion aptly described by former Vancouver *Province* newspaper columnist and Canadian humorist Eric Nicol: "Vancouver citizens half expect their daughters to be violated, but he who lays a rapacious hand on the Park is begging for violence avenging," Nicol wrote in his playful 1970 book of city history, simply named *Vancouver*. The love affair with the park has spread far beyond those who call Vancouver home. In 2014, the millions of users of the travel website TripAdvisor named Stanley Park the best urban park in the world, edging out other heavy-hitters such as Central Park in New York City and Luxembourg Gardens in Paris.

The tranquility and pleasure park-goers experience is interrupted every thirty years or so by fierce fall and winter windstorms that approach from one of two directions—westerlies that sweep down Georgia Strait between the mainland and Vancouver Island and across English Bay, and southeasterlies that take dead aim at the city-facing side of the park and the Stanley Park Causeway. The causeway is a continuation of Highway 99, the main north–south route from downtown Vancouver to North Vancouver, crossing Burrard Inlet on Lions Gate Bridge, a slender, three-lane suspension bridge named after two of the North Coast peaks. The bridge brings into view rugged, snowcapped mountains that lead to more mountains and more mountains, parading into wilderness all the way to the North Pole.

The Columbus Day Storm of 1962 was one of those southeasterly windstorms. It rolled across the US-Canada border late Friday night and earned the title of worst windstorm in history to strike the lower British Columbia mainland. Winds blew steady at 60 miles per hour and gusted to 78 miles per hour. A storm that first unleashed its fury around 5 a.m. off the Northern California coast was still up to no good nearly twenty hours and 1,500 miles later in another unsuspecting zone of death and destruction. Stanley Park was a scene of terror, as falling trees trapped two tour buses and forty-two cars on the causeway. Six cars were smashed by toppled trees.

Renee Archibald, seventy-six, of Victoria, was crushed to death in the back seat of a car driven by her son-in-law, Leslie King, fifty-two, who said a sturdy hemlock some two feet in diameter cracked the car roof as if it was an eggshell. The car had been stalled on the causeway by another fallen tree just before the fatal accident. King, his wife, and his mother-in law had been in North Vancouver visiting relatives and were southbound, returning to the city. "I took one look and knew there was nothing I could do for my mother-in-law," King said the next day from a hospital room at St. Paul's Hospital in Vancouver. He was at the bedside of his wife, Yvonne King, forty-four, one of the six people injured in Stanley Park. She suffered head injuries when the collapsed car roof struck her in the front passenger seat. King dragged his wife from the car and rescue workers used two patrol cars and a stretcher to get her through the maze of fallen trees to the hospital.

Joseph Plag was northbound on the causeway, about to pass the King vehicle. He watched in horror as the tree fell with a roar and crashed on the car. "In another thirty feet, I would have got it, too," Plag told reporter Osmond Turner of the *Province* newspaper in Vancouver. Turner ventured into the park at the peak of the storm, risking life and limb to capture first-hand accounts of the horrific scene. "As he [Plag] spoke, trees exploded and groaned in the darkness around the stalled motorists," Turner described. "The causeway lights went out, flickered on again, and went out as firemen with flashlights hurried by." Turner listened to police officers order the passengers in the tour buses to climb back inside and hug the floors. "The motorists refused, pressing firmly against the outside of the bus," Turner said. "If I'm going to get hit I want to see where it's coming from," one passenger was overheard saying.

Insurance agent Geoffrey Hopkins told Turner that he leaped from his car just as the trees began to fall. His car was surrounded by wrecked cars and fallen branches. Photographer Ray Allan of the *Vancouver Sun* newspaper also walked into the park that night to the scene of the fatal accident. "You'd hear a crack, look up and see a tree swaying, then have to jump as it came crashing down," Allan said. "There was near panic among the people—men, women and children—who had abandoned their cars and were trying to get out of the park." The vehicles trapped on the causeway included an ambulance, two police cars, and two truckloads of cattle. The Lions Gate Bridge, towering two hundred feet above Burrard Inlet, swayed mightily in the wind, prompting the bridge tender to abandon his post and causing highway officials to close the bridge to traffic at midnight. Crews labored through the night to clear fallen trees and limbs from the causeway.

The storm damaged or destroyed more than ten thousand trees in the venerable park. Many were centuries-old conifers that had stories to tell, stories of the First Nations people who lived in the village of Xway-xway in present-day Stanley Park, natives who showered British naval explorer Captain George Vancouver and his fellow explorers with soft white feathers plucked from waterfowl when Vancouver's two ships, in June of 1792, edged into a narrow inlet that separates the park from the north shore of the mainland. Vancouver named the narrow twenty-three-mile-long inlet after a friend and former shipmate, Sir Harry Burrard-Neale. Burrard, who became an admiral in the British Royal Navy, never visited British Columbia, but his name clings to many human endeavors in Vancouver, including a bridge, hotel, and main thoroughfare. Despite the warm reception Vancouver received from the indigenous people, it was almost seventy years before pioneers returned to the Vancouver area.

In the aftermath of the storm, park officials employed thirty to fifty men for several months to buck the fallen trees into manageable lengths, winch them out of the woods, and pile them for sale in the spring. They also felled trees that were damaged in the storm and posed a public safety hazard. But they tried to assure the public that they weren't condoning logging in the park. Park superintendent Stuart Lefeaux insisted the tree removal was not a logging operation. He called it "park forestry." Call it what Lefeaux would, the storm cleanup did not sit well with Vancouverites, noted Nicol. "The cleanup operations of the parks board brought tears to the eyes of the city's old-timers and West Enders," Nicol offered with a hint of hyperbole. "Necessary though it was, it destroyed perhaps forever the wilderness character of much of the Park, the tangled wonderland of rotting giants, brilliant green ferns and saplings springing up willy-nilly amid the bird-filled underbrush."

The seven other direct storm fatalities outside the park, but in the Vancouver area, struck an eerie chord of familiarity with had happened south of the border hours before. They included Raymond Swenns, twenty-two, who was killed when a falling Douglas-fir tree struck his sports car just one-quarter mile from his home in Richmond, BC. Frank Joseph Vleten, twenty-six, died in a two-car collision at a wind- and rain-wracked intersection in Vancouver. Surrey resident Bill Sands, a highway department foreman helping clear trees from fallen power lines along Highway 10 in Surrey, died when a power line suddenly tightened up and tossed a tree through the air that struck Sands.

Toshitaza Richard Koyama, forty-three, drowned at the peak of the storm as he tried to row a small skiff out to his gillnet fishing boat, anchored in Miners Bay off Mayne Island southwest of Vancouver. Accountant Francis Copithorne, fifty-nine, was electrocuted when he stepped on a live power line behind his Vancouver home. Harry Young, seventy, a Vancouver resident and retired firefighter, and Ewald Detter, of Coquitlam, BC, both died of apparent heart attacks while trying to repair storm-damaged television antennas on the roofs of their homes.

Others experienced close calls as the midnight winds howled, the rain fell in sheets, lightning lit up the sky, and downed power lines set off dozens of fires in the greater Vancouver area. In the city's Kitsilano neighborhood on the southwest side of town, Constable Gil Steer was directing traffic around downed power lines when he was struck by a hit-and-run driver, who carried him several feet on the hood of his car before dropping him on the pavement. Steer, a former lineman for the British Columbia Lions professional football team, said the driver got out of his car, looked around, hopped back in his car and drove away. Steer suffered minor injuries to his side and hip.

In another agonizing scene, an unidentified motorist was trapped for an hour on the Fraser Bridge between Lulu Island and Vancouver when a power cable blew down and fell across his car. He started to exit his vehicle, but stopped when someone shouted: "Stay where you are!" The motorist obliged. It was a good thing, because the cable had electrified the bridge, and the unsuspecting motorist would have been electrocuted. Further complicating matters, the bridge tender was not on duty for his midnight shift. He'd been struck in the head by a falling limb as he left home to report to work. The motorist stayed motionless until the power circuit to the bridge was shut off and the bridge was declared safe. It was just another example of life turning on a dime all day and all night along the path of the Columbus Day Storm.

Gale warnings posted in and around Vancouver, British Columbia, helped keep maritime storm damage to a minimum as the Friday night storm pushed into Canada. But at least one Canadian commercial fisherman experienced a brush with death. Louis Kadla spent some ninety minutes in the frigid waters of Vancouver Harbor after his thirty-five-foot trawler, *Norwings*, capsized in Burrard Inlet. He clung to the stern of his boat, and then managed to reach a skiff attached to the fishing vessel. He seized a life ring and started to row for shore some six hundred feet away. But the skiff swamped after just a couple minutes battling the wind-wracked waves. He was back in the water, swimming and fighting for his life.

The fishing boat was discovered overturned in the Indian Arm finger of Burrard Inlet northeast of downtown Vancouver. A city police boat and tug began a search for the vessel's skipper. An hour later, searchers were about to give Kadla up for dead. That's when a shipyard caretaker, William Human, was awakened by his barking dog. Over the crashing waves and howling winds, Human heard a banging sound in a nearby shed. He hastened to the noise and found a bearded man on the floor of the shed, barefoot and soaking wet. Nearly spent and almost unconscious, Kadla didn't remember swimming ashore at Roche Point on the North Vancouver side of Indian Arm. "I remember I climbed over something and the next thing I know for sure is that I was in an ambulance," Kadla told a reporter later Saturday. "I was getting pretty close to the end."

The storm knocked out power to some three hundred thousand BC Hydro electricity customers from Vancouver to Hope, British Columbia, many for more than a week. In the early hours of the blackout, thieves took advantage of shattered storefront windows and darkened streets to steal thousands of dollars of merchandise from downtown Vancouver jewelry, furniture, and dry goods stores. Four of the city's seven radio stations went off the air, as did the two television stations serving a greater Vancouver population of some 770,000 people.

Four days after the storm, the Vancouver *Province* newspaper offered an editorial postscript on the freak of nature:

> As people who usually live climatically cloistered lives, far from the turmoil of typhoons, earthquakes, flood and drought, the residents of our lower mainland reacted like veterans to the havoc of our epic windstorm. And as they repair the damage, they reflect philosophically that, under the law of averages, most of us will never see such a wild night again.

On a drizzly, windless Sunday afternoon in March 2014, I left the cozy confines of the stately Sylvia Hotel, built on the shores of English Bay in 1912, and walked along the seawall promenade. Work on the seawall as an erosion control measure began in 1917 and took more than sixty years to complete. Rain or shine, the seawall pathway is filled with joggers, walkers, bicyclists, baby strollers, and skaters—both city dwellers spilling out of their high-rise condos for exercise and fresh, albeit usually damp, air, and tourists, who add to the some eight million visitors to the park each year.

Ten cargo ships are anchored at odd angles in the bay. If the wind were to blow, they'd all point their bows into it, noted Wolf Read, the Seattle native who in 2015 completed his doctorate studies in forestry science at the University of British Columbia. This kindly, wooly-bear of a guy, equipped with a soft, confident voice, is arguably the foremost expert on the history of Pacific Northwest windstorms. He had agreed to be my tour guide in Stanley Park, showing me the scars and signs of recovery from two noteworthy wind strikes—the 2006 Hanukkah Eve storm and the 1962 Columbus Day Storm. Both storms carried hurricane-force wind gusts through the forests of Stanley Park, each knocking down roughly ten thousand trees from a park inventory of nearly five hundred thousand.

The Hannukah Eve tempest was a westerly wind that tore apart several exposed timber stands, including one at the park's northwest tip—Prospect Point. The Prospect Point area, the highest point in the park, has been replanted, but the saplings have yet to grow tall enough to block the grand views of English Bay and the Georgia Strait generated by the storm. On this sun-forsaken, soggy day, a five-acre patch of the park struck by the Columbus Day Storm is marked by a dense stand of fifty-year-old Douglas-fir. The trees are competing for sunlight and the forest stand could use a thinning to promote tree health. A centuries-old western redcedar tree stands near the trail. The tree's top is missing, as are the tops of many other ancient cedar trees we saw in the park. It's safe to bet some were victims of the 1962 windstorm; others may just be showing their age. A sign posted on the tree trunk tells visitors that some of the Douglas-fir trees will be selectively removed to allow the remaining trees to grow with more sunlight and vigor.

It's a project supported by Read's forestry professors at UBC, but not necessarily appreciated or supported by the general public, which didn't like the sound of chain saws in the park in 1962, and still doesn't. But even an untrained eye can tell the trees' growth is stunted as they compete with each other for sunlight and nutrients in the soils. The shady forest floor supports prickly salmonberry bushes and very little else. We caught glimpses of cars and trucks through the trees, and heard their tires whooshing across the rain-slickened pavement, reminding us how close to the Stanley Park Causeway we stood. My mind filled with images of unsuspecting motorists trapped in the park the night of the storm, fearing for their lives.

The next day I entered the park on foot again, this time along the shores of Coal Harbour, home to the venerable Vancouver Rowing Club and its 1911 Tudor-style clubhouse. The rower's clubhouse separates the southern end of Stanley Park from a cityscape filled today with some seven hundred concrete and glass skyscrapers, many of them skinny, pale-green, and gray and owned in part by mainland Chinese moneymakers and Hong Kong immigrants, in a city where 25 percent of the inhabitants speak Chinese as their first language. In 2014, Vancouver ranked ninth on the world's list of cities with the most high-rise buildings, a rather remarkable ascension for one of the world's youngest cities, one that incorporated on April 2, 1886, and burned to the ground just two months later. Following a 1956 rezone of the downtown area to accommodate rapid vertical growth, the city began to sprout tall buildings like mushrooms.

I ate lunch at Stanley's Bar and Grill, housed in the Stanley Park Pavilion, also built in 1911. I washed down my chicken sandwich with a Stanley Park Windstorm Pale Ale, hoisting my glass to the memory that drinking alcohol in public eateries in Vancouver wasn't allowed until 1952. Our waitress told me that sixty cents from every Windstorm Pale Ale sold is donated to the Stanley Park Ecological Society to support park restoration efforts. Drinking beer to enhance the park's future: it doesn't get any better than that.

At 1 a.m., Saturday, Capell and his KGW news team entered their ninth hour of continuous storm coverage back in Portland, Oregon. Capell shared with his listeners the latest wind readings he had gathered from the weather bureau: 10–20 miles per hour in Portland, 10 miles per hour in Eugene, 35–45 miles per hour in Seattle, 92 miles per hour in Bellingham, and 56 miles per hour in Victoria, British Columbia. Clearly the storm had moved north, and the Portland area was in the clear. "There's no sign of any reoccurrence," he said, a hint of relief and weariness in his voice. "This is the most encouraging report we've had. The winds have dropped very significantly to the south of us. It looks as if this big windstorm of Friday night is finally moving to the north."

At 1:19 a.m., Capell and the KGW news team signed off the air. Capell drove home to his wife and infant son, and a house with part of its roof blown off, a shattered storm door window, a lawn strewn with shingles, a badly damaged fence, and an unhinged plum tree in the backyard that he would prop up, giving it a new lease on life.

15
A Stormy Aftermath

After shredding Vancouver, British Columbia, the storm pushed north and slightly west, grazing the east coast of Vancouver Island in the early morning hours Saturday. The storm's low-pressure center filled in rapidly as it crossed the cool waters of Georgia Strait and ran into the rugged, coastal mountain terrain. The winds slowly subsided from a scream to a roar to a growl to a sob to a sigh. A full twenty-four hours after the storm had intensified off the Northern California coast and began its punishing march north, it finally washed out to nothing more than a typical winter storm with unsettled winds and a soaking rain.

The winds died down and the storm front disassembled. But a third storm lashed the Northern California area Saturday. It marked the third straight day of heavy rains in the Bay Area, including the greatest twenty-four-hour October rainfall in California history: 14.10 inches in the Santa Cruz Mountains twenty miles south of San Francisco October 12–13. The village of Orinda near Oakland recorded a staggering 18.41 inches in two days. By comparison, rainfall in Oregon and Washington October 11–13 was nothing exceptional, including Salem, 1.62 inches; Portland, 1.12 inches; Olympia, 0.33 inches; and Seattle, 1.66 inches.

The pulverizing Bay Area rains had by Saturday turned Oakland's residential hillsides into oozing torrents of mud and stormwater, sweeping away homes, forcing hundreds of evacuations, and blocking both the Eastshore and Nimitz Freeways. City officials described it as the worst storm in the city's history. It was a tragic one, too, burying two children alive.

The first mudslide fatality happened at 8:30 a.m., Saturday, in the upscale Crestmont district in Oakland's eastern hills. A waterlogged embankment collapsed above the home of Mr. and Mrs. James Dobson and their two young children, James Jr., two, and Diane, five. The mother heard the rumble of an avalanche descending on their home. She gathered the boy in her arms and grabbed her daughter's hand and ran across the living room in a futile dash for safety. But a wall of mud, rocks, and water crushed the garage above the home and shoved the family car through the roof of the house, opening a pathway for debris and mud to violate the home and rip a daughter's hand from a mother's grasp. The daughter was buried in the debris. The mother escaped with a broken leg and James Jr. was unharmed.

Bill McCarthy and his eight-year-old son, William, were sandbagging a retaining wall behind their home in Orinda Saturday morning when the wall collapsed, drowning the boy in an avalanche of mud and water. A mother and daughter were buried alive in a mudslide in Oakland Saturday when their car slid off an embankment and slowly sank as witnesses watched in horror. Then the car horn honked, triggering a rescue effort that freed the two after seventy frantic minutes. "I'm certainly glad to see you," the rescued woman said. Then she fainted.

Tragedy was not over yet in a Bay Area region that tallied some 450 homes destroyed or damaged by mudslides and flooding. On Sunday, two-year-old Diana Terry of San Jose fell into a water-filled hole in a neighbor's yard and drowned. And San Francisco physician Dr. Lloyd F. Quirin, forty-four, succumbed Sunday to injuries he received when his car skidded out of control on Highway 101 in Marin County Thursday night during the first wave of the storm.

Sunday was also the day that play resumed in the 1962 World Series between the San Francisco Giants and the New York Yankees, two fabled baseball teams with seven future Hall of Famers on the rosters, include the Yankee's Mickey Mantle, Whitey Ford, and Yogi Berra, and the Giant's Willie Mays, Willie McCovey, Juan Marichal, and Orlando Cepeda. The fall classic, led by the Yankees three games to two, had already experienced a rare rain-out in New York, pushing game 5 back to October 10. The Yankees won that game on a three-run home run by baseball's Rookie of the Year Tom Tresh. Game 6 had originally been scheduled for Thursday, October 11, at Candlestick Park in San Francisco, but the storm made mincemeat of the schedule.

Some of the players groused about the storm delays. Others took the foul weather in stride. "It can't hurt my batting eye," Yankee slugger

Mantle joked Saturday with *San Francisco Chronicle* sportswriter Art Rosenbaum. Mantle had just two hits and no home runs in the first five games of the series.

The Giants won the rescheduled Sunday game 6 by a five-to-two score, thanks to Cepeda's three-for-four day at the plate. That set up game 7 the next day. Taking advantage of the storm delays, Yankee skipper Ralph Houk sent game 5 victor and twenty-three-game regular-season winner Ralph Terry to the mound for the third time in the series. It turned into a classic pitcher's duel. Terry hurled a perfect game until the sixth inning and a two-hit shutout into the ninth. The Giant's Jack Sanford was almost as good, allowing just one run in the fifth.

Giant's pinch hitter Matty Alou opened the ninth with a bunt single, but his brother, Felipe Alou, and Tom Haller, struck out. That brought Mays to the plate. He lined a ball down the right field line. Suddenly, the storm was back in play: the soggy field from the heavy rains slowed the ball down, giving right fielder Roger Maris—an underrated defensive player—enough time to get to the ball before it could slip into the corner. In a decision second-guessed for months, Giant's third-base coach Whitey Lockman held Alou at third and Mays coasted into second with a double. McCovey came to the plate. He smashed a line drive, but right at second baseman Bobby Richardson, and the Yankees had their twentieth World Series title in team history.

To the north, the college football game in Portland, Oregon, slated for Saturday afternoon between Oregon State and Washington was very much in doubt. The Friday night storm blew four sections of roofing off Multnomah Stadium's north end zone and another hundred-foot section of roofing off near the press box. The television section of the press box was ripped loose by the wind and temporary bleachers were sent flying.

One would think that the stadium damage, combined with widespread power outages and general chaos in the Portland area, would force cancellation of the game. But school officials surveyed the damage early Saturday morning and made the call at 9 a.m. to play the game as scheduled. Roofing and other debris littered the field Saturday morning. Crews worked feverishly to clean up the mess. The cleanup continued right up to the 1:30 p.m. kick-off, and storm debris was still piled next to the field at game time. The decision to play amid the storm wreckage speaks to the priority of the day—sports trumped common sense and the inconvenience caused by the worst windstorm in Pacific Northwest history.

The stadium scoreboard lacked electricity, and the teams had to don their uniforms by candlelight in the Multnomah Athletic Club. But the sun shined bright on the soggy field. Some 30,030 resolute fans braved the power outages, downed trees, tangled power lines, and storm damage to their homes to attend the game. It turned into a hard-fought contest featuring OSU's triple-threat quarterback Terry Baker and the University of Washington Huskies, ranked seventh in the Associated Press college football poll and two years removed from their seventeen-to-seven Rose Bowl victory over the then-number-one Minnesota Gophers. The dynamic passing duo of Baker to Vern Burke (seven catches, 103 yards) wasn't quite enough. The Huskies took advantage of two late-game Beaver turnovers, and a two-yard run by halfback Charlie Mitchell late in the fourth quarter, to eke out a fourteen-to-thirteen victory.

After the game, both teams were treated to icy showers in cold dark locker rooms that resembled caves dimly lit by flashlights and lanterns. "Who the hell cares!" one shivering but exuberant Huskies ballplayer shouted as he exited the shower. Baker stayed in Portland for the weekend, as did other Portland hometown teammates. They headed to a party at teammate Rich Brooks's parents' home in Lake Oswego, the heavily wooded Portland suburb that suffered so much tree damage the night before. Navigating the fallen trees and downed power lines was next to impossible, Baker recalled.

The UW team went on to a seven-one-two season and finished fifteenth in the United Press International coach's poll, but wasn't invited to one of the ten bowl games. The 2013–2014 season, by comparison, saw an excessive thirty-five postseason games. OSU finished nine-and-two in 1962 and earned an invitation to play Villanova in the Liberty Bowl December 15 in Philadelphia, a game the Beavers won six to nothing on a ninety-nine-yard, first-quarter touchdown run by Baker.

In 1962, Baker won college football's coveted Heisman Trophy, the first player so awarded west of Texas. He also earned the Maxwell Award and the Sports Illustrated Sportsman of the Year award. Most of his many sports trophies were donated to the OSU sports department. The Heisman Trophy eventually became a doorstop in a seldom-used bedroom at his Portland area home.

Portland General Electric crews worked around the clock to restore power in the days following the Columbus Day Storm. Courtesy of the Oregon Historical Society.

Storm-battered residents along the storm path from Southern Oregon to British Columbia woke up Saturday to mostly sunny skies and mild temperatures, in sharp contrast to their rain-soaked neighbors to the south. The benign weather was an irony of sorts, framed against neighborhoods strewn with storm debris and crippled by widespread power outages affecting more than 1.3 million electricity customers. In some rural areas of the region, it took two or more weeks to restore power. Retired Portland General Electric lineman R. J. Brown recalled working on a tangled mess of broken poles, snapped cross beams, and burned-out transformers. "Ninety-eight percent of the system was down," Brown said. "That hadn't happened before [the storm] and it hasn't happened since."

Longview, Washington, had some three hundred miles of lines lying on the ground or dangling in the air. On the outskirts of Portland, forty heavy-duty poles carrying a 115,000-volt transmission line had to be replaced. The electric grid was like a dismantled jigsaw puzzle that had to be pieced back together. "In some places we can hook up one line and restore 30 homes," one power official said at the time. "In others we have to lift 30 trees to get one house back."

Cathy Schoenborn McLean was a young girl living in Independence, Oregon, in 1962. Her father, Milo Schoenborn, worked for Pacific Power and Light. She didn't see much of her dad in the days following the storm. "They worked until Monday night and came home for only six hours," she said. "The next stint was for thirty-two hours with six hours off. The rest of the time they worked eighteen hours and had six off. The men were like zombies." Utilities from as far away as Montana and Wyoming dispatched crews and equipment to help rebuild the storm-stricken power grid. Five days after the storm, PP&L placed an advertisement in newspapers read by its two hundred thousand customers, stretching three hundred miles from Northern California to the Columbia River, thanking them for their patience and vowing to have power to all customers restored soon. Company officials said, "We've been digging our way out of the biggest and most difficult electric service problem ever experienced in the West!" Some forty-five thousand Oregon families were still without power, and thirty-five thousand telephone customers lacked phone service for a week following the storm's passage.

Homeowners and city public works crews also set out to clean up the storm debris. In Corvallis, Oregon, Oregon State University forestry

Exhausted electric utility workers grabbed a little rest whenever they could in the around-the-clock effort to restore power in the wake of the storm. Courtesy of the Oregon Historical Society.

Students pitched in to help
clear the storm debris on the
Oregon State University campus
in Corvallis. Courtesy of the
Oregon State University Valley
Library Special Collections and
Archive Research Center.

students formed work crews in the aftermath of the storm. They put their
experience with chain saws to work clearing trees and limbs all over the
heavily wooded college campus. After working about five days on campus,
the crews extended their cleanup work to city streets and neighborhoods.
Hundreds of other college students pitched in to collect the storm debris
and load it into trucks for disposal.

All the downed trees and limbs touched off a chain-saw-buying frenzy.
McCullough Corporation shipped 1,100 chain saws to the Pacific Northwest in the first five days after the storm, equal to a normal year's business
for chain saw wholesalers and retailers in the region. Cities along the storm
path set up debris drop-off centers that burned for months after the storm.
"For a time the sound of chain saws could be heard all over the campus and
Corvallis," said Florence Turner Gerlach, a Portland resident who was an
OSU student living on campus in 1962. She felt the storm robbed the college campus of much of its beauty and charm, especially the Quadrangle,
which was home to many magnificent old trees that were knocked down
by the hurricane-force winds.

Tree losses on college campuses and city parks played out all along the
path of the storm. A few blocks from Tacoma General Hospital, where doctors patched up Charley Brammer, the winds whistled through Tacoma's
historic Wright Park, a twenty-seven-acre park donated to the city in 1886

by one of the mainstay boosters of the "City of Destiny," Charles B. Wright. Trees from four continents were planted there to create an elegant arboretum covering ten city blocks. The storm knocked down or ruined about 15 percent of the park's more than eight hundred trees. City residents flocked to the park Saturday morning after the storm, and huddled in groups and wept as crews went to work on the tree carnage with chain saws.

Not all the historic losses were trees. In the Willamette Valley east of Salem, the storm destroyed a century-old barn on the old Nicholas Shrum homestead. In 1860—one year after Oregon achieved statehood—six pro-slavery Democratic state legislators hid in the barn for eleven days in a failed attempt to avoid a quorum needed to elect two anti-slavery US senators.

Claims adjusters in Oregon and Washington went right to work, processing nearly one hundred thousand damage claims totaling more than $24 million in 1962 dollars. Pioneer Mutual Insurance Company of Hillsboro, Oregon, was the oldest mutual fire insurance company in Oregon, founded in 1883. Just before Christmas, ten weeks after the storm, the company was placed in receivership by a rush of claims and liabilities that outpaced its assets by more than $131,000. "Pioneer Mutual was a fine old company," remarked Oregon Insurance Commissioner Walter Karlann. "It was in fine financial condition before the storm."

Oregon governor Mark Hatfield telegraphed President John F. Kennedy October 13, making the case for disaster relief funds from the federal government to cover 10 percent of the damage to public agencies, including cities, counties, and schools. The governor requested $2.7 million (about $21 million in 2016 dollars). A preoccupied president authorized an initial release of $1 million ($7.9 million in 2016 dollars) on October 22, the same day he addressed the nation about the Cuban Missile Crisis. Capell's KGW television station did not resume broadcasting until Tuesday, October 16, the same day power was restored to the Capell home. Capell spent sixteen hours at the station that day as the news team for the NBC affiliate put together special storm coverage for a hungry television audience tuned in to Channel 8.

The deadly storm lingered on the front pages of Pacific Northwest newspapers for several days, but soon took a back seat to coverage of the Cuban Missile Crisis. Residents of the nation's far northwest corner didn't even have time to process the trauma from the storm or repair the damage it caused before they were confronted with this new existential threat to their safety.

Epilogue

Life was not kind to many of those who played significant roles in the storm. Capell, the unsung World War II veteran and popular Portland broadcast meteorologist, was called to battle once again ten years after the storm. This time the enemy was primary lateral sclerosis, a type of motor neuron disease akin to amyotrophic lateral sclerosis, commonly known as Lou Gehrig's disease. Primary lateral sclerosis (PLS) causes weakness in voluntary muscles, the ones that control a person's legs, arms, and tongue. PLS progresses more slowly than ALS and, in most cases, isn't fatal.

Despite a debilitating condition that weakened his muscles and slowly led to paralysis, Capell stayed on the air at KGW for forty-four years. He also continued to serve as the public address announcer at Portland Buckaroo hockey games, sailed a boat he moored at a Columbia River marina, and played tennis with his longtime coworker at KGW, sportscaster Doug LaMear. "If that had been me with PLS, I probably would have checked out, or become a drunk," LaMear reflected years later. "Jack was a super guy and a World War II hero who should have received a medal."

In the later years, as Capell's mobility deteriorated from the disease, his wife—later aided by their oldest son, John—would take him to the station, get him seated on the set, and write the temperatures on the in-studio weather board, in an era before computer-aided weather maps and reports. After he delivered his weather report, they would help him up from the studio desk and take him home. "It was a lesson in love and dedication," said Floyd McKay, the journalist who covered the Columbus Day Storm for the *Oregon Statesman* newspaper in Salem, Oregon, then joined the KGW news team shortly before Capell was diagnosed with his debilitating disease. "Until the day he left, there were many, many Portlanders who still waited on Jack's word for the weather, even when he broadcast only on Sunday." Capell stayed on the air until September 3, 2000, still enthused about a long fulfilling life as a weather forecaster, a craft he considered a science, but also a game. Thunderstorms rumbled through the Portland area the night of Capell's last weather forecast.

After he retired, Capell received the National Academy of Television Arts and Sciences' Silver Circle Award for lifetime achievement in broadcasting. He was also inducted into the West Seattle High School Hall of Fame in 2001. Capell and his wife, Sylvia, moved back to Seattle and lived in a home overlooking Puget Sound. She died in 2003. The Portland television pioneer, World War II hero, die-hard hockey fan, and devoted family man died in Seattle on June 14, 2009. He was eighty-six.

John Capell, a television news producer based in Portland, Oregon, offered this fine tribute at his father's memorial service, June 27, 2009, in Seattle: "I think of the courage and the strength of character when you realize the last forty years of your life, your body betrayed you and you can't do what you want with your body, and it doesn't get you down," his oldest son said. "He gave me a lot of gifts—integrity, courage, honesty—those are gifts he gave without saying or doing. He just gave by example."

Female aviation pioneer Ruth Wikander was flying that strange contraption called an Aerocar the day of the violent windstorm because Friday was a day off for the car-plane's regular pilot, Wayne Nutsch, whose radio name in the air was "Scotty Wright." Nutsch, an Independence, Oregon–based aviation consultant, said that Wikander, a dear friend who never married, died in an airplane crash on February 11, 1968, near Wilsonville, Oregon. She was the flight instructor in a twin-engine Beechcraft B95, giving single-engine instruction to a student pilot when the employed engine failed or malfunctioned, sending the plane into a fatal stall and spin with the student at the controls, according to the National Traffic Safety Board accident report. The student pilot and a passenger also died in the crash.

Wikander, who had more than 10,000 hours of flight time, including 121 hours in the Beechcraft B95, couldn't intervene because the plane was not equipped with a second set of controls. All she had was a throw-over yoke, one set of controls that could be operated by only one pilot at a time, but from either side of the front seat. Under today's Federal Aviation Administration rules, most flight instruction in single-control aircraft is prohibited.

After one year and one thousand hours flying over the streets of Portland, Aerocar N103D was purchased by a Mossyrock, Washington, man who worked as a traveling salesperson for Procter & Gamble. He used it as both an airplane and a car, with the detached wings and tail towed behind. Taylor, the plane's designer-builder, reacquired the car-plane after it was blown off the road into a ravine during a fierce South Dakota windstorm. Taylor rebuilt it, and the hybrid car-plane changed hands several times, flying for the last time in 1977. It has been sitting in a garage in Grand Junction, Colorado, since 1981, with an asking price of $2.2 million. Taylor, a pilot who worked on the navy missile program during World War II, tried for years to bring his invention to mass production. He came close in 1961, and again in 1970 when Ford Motor Company showed interest. But an oil crisis and increased imports of Japanese cars dampened the auto giant's interest. Taylor, and his dream of a mass-produced flying car, died in 1995. He was eight-three.

Called by many the cultural and historical conscience of Portland, Francis Murnane and his civic good deeds came to an abrupt halt April 10, 1968, when he died at age fifty-three of a heart attack while presiding over a meeting at the ILWU Local 8 union hall in downtown Portland. Included on the agenda was a vote to grant full union membership to Alt McNeil, a black man, who was among the first African Americans to break the Local 8 color barrier in 1964. Before the successful vote, Murnane asked for a moment of silence. The Reverend Martin Luther King Jr. had been assassinated six days earlier. Murnane's funeral drew more than five hundred people, including good friend and honorary pallbearer ILWU International president Harry Bridges, Governor Tom McCall, US senator Wayne Morse, and Portland mayor Terry Schrunk. The Port of Portland shut down operations that day so longshore workers could attend their fallen brother's Roman Catholic last rites.

Ten years after his death, the Portland waterfront that was Murnane's second home said goodbye to the four-lane Harbor Drive along the shores of the Willamette River and said hello four years later to a thirty-seven-acre riverfront park, which was named in 1984 after former Oregon governor Tom McCall, Capell's former colleague in the KGW newsroom at the time of the storm. Union leaders had urged the city to name the waterfront park after Murnane. The city instead installed a working wharf and plaque in Murnane's honor just south of the Burnside Bridge at the site of Portland's first commercial wharf, the Waymire Dock. The plaque was stolen in 1989, then replaced. The wharf was condemned as a safety hazard in 2006. Sadly, the only memorial to recognize a Portland labor leader was destroyed in 2010 to make way for the expansion of Portland's Saturday Market.

An unofficial review of the weather bureau's forecast one year after the storm bluntly stated that the federal agency was negligent for not updating the morning forecast later in the afternoon after reports of 86-mile-per-hour winds in Eugene at 4 p.m., and 70-mile-per-hour winds in Salem at 4:20 p.m., reached the Portland office on October 12. The critical report was written by Jim Holcomb and Glen Boire, two of the weather bureau forecasters working in the Portland office the day the storm struck. "This would have given about an hour and a half notice to the public and other interests, and might have been beneficial," the two forecasters wrote.

Holcomb retired from the National Weather Service in 1989 after a thirty-year career with the federal agency, much of it in Wenatchee, Washington, where he continues to live and work, providing weather forecasting services for commercial fruit growers on the east side of the Cascades. He pulls no punches when he looks back on what happened the afternoon of October 12, 1962: "The reports from Eugene and Salem came in right at shift change. I wanted to get home. I feared for the safety of my wife and six-month-old son," he said. "We should have put out a 'now-cast' [a real-time update of a previous weather forecast]. It was my fault we didn't do it." The postmortem of the Columbus Day Storm forecast was never accepted or published by Holcomb's superiors in the weather bureau. It was viewed as too critical of the agency. Other criticisms of the weather bureau were voiced in the days after the storm, including complaints and finger-pointing from school, highway, and local government officials who said they didn't receive advance notice of the storm.

The high wind warning issued 10:40 a.m. Friday morning by the Portland weather bureau office was widely distributed to media outlets for immediate and frequent broadcast throughout Western Oregon. In addition, some two dozen utilities, airports, road departments, and municipal and school contacts in the greater Portland area were contacted by weather forecasters. However, the wind warning slipped through cracks in an emergency preparedness system more attuned to Cold War threats than natural disasters. A case in point is the Oregon State Highway Department investigation by emergency planning engineer John F. Hagemann into why the storm caught state highway officials off guard.

The department's regular teletype connection with the weather bureau ran from November 13 to April 15, so it was not operating when the storm struck. Highway officials relied on the Oregon State Police in Salem for weather warnings prior to the official storm season. The police desk sergeant interviewed by Hagemann said the police desk was working off a 9 a.m. wind warning, from the US Weather Bureau office at McNary Field in Salem, which called for afternoon winds 20 to 30 miles per hour with gusts to 50 miles per hour. The sergeant was probably referring to the wind warning Salem weather officials received from their counterparts in Portland that called for winds 30 to 40 miles per hour in the late afternoon with gusts to 60 miles per hour or more. The warning was telephoned to the state police office for distribution to all state agencies, according to Salem weather records. "The desk sergeant further stated that, as this was not an unusual weather condition, no 'warning' was transmitted to other state agencies," Hagemann said in a memo to assistant state highway engineer Tom Edwards. "It would appear that no Highway employee had any official advance notification of the seriousness of the storm."

All along the path of the storm, some schools dismissed their student in advance of the storm while others seemed oblivious to the storm and its might. "I'm not blaming the weather bureau in this situation," Salem schools superintendent Charles Schmidt said at a post-storm school board meeting. "I just want to get the record straight. No one contacted our office at the school administration building Friday." By the time forecasters in the Salem weather bureau warned of winds that could gust to 80 miles per hour, it was 4 p.m., classes had been dismissed, and the winds were just minutes away. The highest surface wind gust registered at the Salem airport was 90 miles per hour at 5:58 p.m.

Capell and others defended the actions of the federal agency and its forecasters, saying they did the best they could with the scant information they had. "A careful review of the events of the Terrible Twelfth leads to the inescapable conclusion that rather than be criticized, the Portland forecasters who faced the meteorological problems of that most difficult day should be praised for their excellent judgment," Capell wrote.

The civil defense system also came under fire in the days following the storm. The federal program, which grew in the 1950s as the threat of nuclear war grew, too, had a presence in most American communities and was synonymous with "drop-and-cover" exercises in public schools, the creation of a network of fallout shelters, and other futile efforts to prepare in case of a nuclear exchange with the Russians. Civil defense programs received very little political or financial support at the local, state, or federal level, but they were the face of emergency management in the United States. That began to change in 1961, however, when President John F. Kennedy created the Office of Emergency Preparedness inside the White House, ostensibly to manage and respond to the risk of natural disasters.

"If Civil Defense can't come forward to do something in an emergency like this, when on earth will they?" Eugene mayor Edward Cone barked at an October 16 meeting of Lane County mayors, four days after the Columbus Day Storm. "We are not miracle workers," responded James Koepke, director of a county civil defense program consisting of him and two part-time employees. He said most Lane County cities were contacted by his office during the storm, but none asked for help. Koepke said the storm did teach some valuable lessons: citizens need to equip their home with emergency food supplies, flashlights, water, and battery-powered radios. And more hospitals and radio stations need generators to produce electricity during widespread power outages.

The Columbus Day Storm was one of several natural disasters in the 1960s that prompted a new federal response to emergency planning. The Ash Wednesday storm of March 6–8, 1962, featured storm surges that devastated more than 620 miles of East Coast shoreline in six states and killing forty people. Seven months later, the Columbus Day Storm struck. Then, in 1964, the Great Alaskan earthquake created another kind of chaos and upheaval. The temblor, which measured 9.2 on the Richter scale, and the subsequent tsunamis it generated as far south as Northern California, claimed 123 lives. Deadly Gulf Coast hurricanes—Betsy in 1965 and Camille in 1969—killed hundreds more and added to the growing reality that natural disasters begged for a more coordinated federal, state, and local response. This led to the creation of a national flood insurance program in 1968; the 1974 Disaster Relief Act, which provides funding for disaster response, emergency housing, and other assistance to stricken communities; and the Federal Emergency Management Agency in 1978.

Predicting regional storm patterns in an uncertain climate future is no easy task. But, meteorologists today are much better equipped to detect in real time with a nontropical windstorm bearing down on the Pacific Northwest off the Pacific Ocean. A typhoon such as Typhoon Freda would not sneak up on the West Coast. It would be tracked by weather satellites, ocean weather buoys, and aircraft all the way across the Pacific Ocean, and watched closely to see if it mixed in an explosive way with other high- and low-pressure systems before it made landfall.

"We would have anywhere from three to five days of advance notice," said Ted Buehner, Seattle-based National Weather Service warning meteorologist and a Columbus Day Storm survivor. "Since the 1990s, we've predicted every major windstorm days before they arrived," Mass said on the fifty-year anniversary of the Columbus Day Storm. "That's a tremendous accomplishment."

That doesn't mean they have a perfect batting record predicting the intensity of the windstorms. A case in point: Typhoon Songda, which reached maximum strength southeast of Japan on October, 11, 2016, with winds peaking at 150 miles per hour, showed up in long-range weather forecasts as a threat to sweep across the Pacific Ocean in a way eerily similar to Typhoon Freda of Columbus Day Storm fame fifty-four years earlier, if it made the successful transition from a tropical typhoon to a midlatitude cyclone. If the storm's low-pressure center traveled close to shore in the Pacific Northwest, it could rival or supplant the Columbus Day Storm's destructive legacy, the weather computer models suggested of a storm expected to strike the region on Saturday, October 15.

The UW's Mass sounded the alarm for the advancing windstorm on October 11 on his popular weather blog, calling it a "true monster storm, potentially as strong as the most powerful storm in Northwest history (the Columbus Day Storm of 1962)." However, he cautioned that there was a lot of uncertainty about the intensity and path, for a storm that has to cross the entire Pacific Ocean and transition from a tropical storm to a midlatitude cyclone. The National Weather Service issued an alert on October 13, one day after the anniversary of the Columbus Day Storm, suggesting a one-in-three chance that the approaching extratropical storm likely to be born from Typhoon Songda could be one for the ages and severely damaging to Western Washington. But the weather forecasting models also showed a higher likelihood that two smaller windstorms would make landfall Friday and Saturday. They would be powerful, but more like the windstorms the region sees almost every winter.

The mere mention in the mass media and social media late into the week of a storm potentially as strong as the Columbus Day Storm caused a public rush on emergency supplies such as bottled water, batteries, and generators. Schools and nonprofit groups in the Puget Sound region cancelled events scheduled for Friday night and Saturday. This reaction speaks at least in part to the stature of the Columbus Day Storm, especially in the minds of those who lived through it. Every time a typhoon bears down on the Pacific Northwest and morphs into an extratropical cyclone, the memory of the mother of all storms roars back to life, reminding us that, yes, we are vulnerable to deadly, damaging windstorms and, yes, the possibility of more severe weather events as the earth continues to warm can't be ignored in the face of an uncertain climate future.

The storms did not live up to their worst-case scenario billing, although the Friday windstorm spawned a freakish tornado that touched down in Manzanita on the Oregon coast, damaging several buildings. The Oregon and Washington coasts were lashed by winds gusting 40 to 70 miles per hour. Winds struggled to reach 40 miles per hour in Seattle Saturday evening. The low-pressure center for the Saturday storm was a full sixty miles west of the worse-case scenario predictions, far enough offshore to limit damage in Puget Sound.

In a storm postmortem Mass posted on his weather blog October 16, he suggested that the weather forecasting shortcomings of the previous few days were not so much a failure of the models as they were the failure to communicate the uncertainty inherent in those models. "Local meteorologists warned of the worst-case scenario, but failed to communicate the uncertainty of the prediction," he said. "I tried to talk about [storm] track errors, but it is clear that I needed to do much more." Some of the forecasting uncertainty could be reduced if a weather radar station was placed on the Central Oregon coast to fill in a gap in the system, Mass said. Also, a high resolution ensemble forecast system for the entire United States would make it easier for meteorologists to do their jobs, he said.

A Columbus Day Storm repeat today would assault a far different landscape than the one that existed in 1962. The region's population has more than doubled—more than 5.25 million people in Western Washington in 2012, compared with 2.2 million in 1962. The 1960 census for the Portland metropolitan area, which encompasses six storm-damaged counties, was slightly more than 875,000. By 2015, it was close to 2.35 million. "We have more people and things in harm's way than we did fifty years ago," said weather warning meteorologist Ted Buehner. "We are so dependent on power and technology today. Cell phone towers would go down in the wind—there would be a lot of crippling power and communication breakdowns."

Meteorologists have long puzzled over the Columbus Day Storm—its violent development, its stubborn persistence, its mighty reach, and its destructive power. They've figured out some of the storm's equation, but not all of it. And that's okay. The storm is an outlier, and outliers imply a sense of mystery. They reside in a place where not all is explained. And so is it with the Columbus Day Storm, a powerful act of nature that allowed for, no, insisted on, gaps in knowledge, unanswered questions, and a cloud of mystery over a stormy world partly understood and partly left for us to ponder.

Appendix

SAFFIR-SIMPSON HURRICANE WIND SCALE

A ranking system for hurricanes and typhoons was created in 1974 by two Americans, Herbert Saffir (1917–2007), a civil engineer, and Robert H. Simpson (1912–2014), a meteorologist and first director of the National Hurricane Research Project.

CATEGORY 1: Winds 74–95 miles per hour, barometric pressure greater than 28.94, dangerous winds that can damage roofs, siding, and gutters of homes, knock down shallow-rooted trees, snap large branches and power lines, and trigger power outages lasting up to several days.

CATEGORY 2: Winds 96–110 miles per hour, barometric pressure 28.50--28.91, extremely dangerous winds that can cause major damage to the roofs and sidings of homes, mobile homes, signs, and boat moorages, uproot and break many trees, block roads, and cause power outages lasting several days to weeks.

CATEGORY 3: Winds 111–130 miles per hour, barometric pressure 27.91--28.47, devastating damage to well-built homes, many fallen trees, power lines, and blocked roads, electricity and water service could be disrupted several days to weeks.

CATEGORY 4: Winds 131–155 miles per hour, barometric pressure 27.17--27.88, catastrophic damage to buildings, widespread tree damage and road blockages that isolate neighborhoods, some areas uninhabitable, utility outages could last weeks to months.

CATEGORY 5: Winds greater than 155 miles per hour, barometric pressure less than 27.17, catastrophic damage, many roof and wall failures, most trees and power lines down, much of the area uninhabitable for weeks or months, utility outages lasting weeks to months.

Sources: National Hurricane Center of the National Oceanographic and Atmospheric Administration; Scott Huler, *Defining the Wind* (New York: Three Rivers Press, 2004), 258–259; Jack Williams, *The Weather Book* (New York: Vintage Books, 1997), 145.

Source Notes

CHAPTER 1: OUT ON A LIMB

1   "Jack Capell opened . . . " Jack Capell's historic weather forecast and return trip to the KGW station from the US Weather Bureau office comes from Stuart Tomlinson, "Predictor of 'Storm of the Century' Dies," *Oregonian* (Portland), June 16, 2009, B-2; author interview with Capell in October 2002 and interviews with his sons, John and Tom Capell, and his former KGW colleagues, Doug LaMear and Floyd McKay, in the summer of 2015.

2   "The thirty-nine-year-old weather forecaster . . . " Capell's early years and World War II experiences are chronicled in Jack Capell, *Surviving the Odds* (Claremont, CA: Regina Books, 2007).

2   "World War II casualties . . . " Fourth Infantry Division casualty from University of Washington history professor Jon Bridgman, in his Foreword of Jack Capell, *Surviving the Odds* (Claremont, CA: Regina Books, 2007), 12.

4   "He entered a newsroom . . . " A rebroadcast of the KGW AM 620 coverage of the Columbus Day Storm is available at, "Portland Radio History," clip no. 7, pdxradio.com.

5   "Priscilla Julien and a friend . . . " Priscilla Julien letter submitted May 17, 1963, Columbus Day Storm Collection, manuscripts 1038, Oregon Historical Society Research Library, Portland, Oregon.

5   "Mary McDonald Eden was . . . " response submitted by Eden on August 15, 2004, to Vancouver *Columbian*, Vancouver, Washington (www.columbian. com/history/columbusdaystorm/).

6   "Our advice is to stay inside . . . " KGW AM 620 rebroadcast of the storm coverage.

6   "It's packed like a can of worms . . . " From "Crowds Jam City Hotels," *Oregonian*, October 13, 1962, A-12.

6   "The winds were strong enough . . . " Fred W. Decker, Owen P. Cramer, and Byron P. Harper, "The Columbus Day 'Big Blow' in Oregon," *Weatherwise*, December 1962, 214.

6   "They were built to withstand . . . " Dorothy Franklin, *West Coast Disaster: Columbus Day 1962* (Portland: Gann Publishing, 1962), 124.

7   "All five Portland television stations . . . " "TV Stations Knocked Out," *Oregonian*, October 14, 1962, A-13.

7   "Transistor radios, which had just . . . " Simon Winchester, *Pacific* (New York: HarperCollins, 2015), 108–109.

CHAPTER 2: TRACKING TYPHOON FREDA

9   "The developing storm's first breath . . . " Birth of storm noted in *Annual Typhoon Report 1962*, issued by US Fleet Weather Central/Joint Typhoon Warning Center (San Francisco, California), January 28, 1963, 178 (see at metoc.ndbc.noaa.gov/JWTC/annual-tropical-cyclone-reports). The name of Eniwetok Atoll was changed to Enewetak Atoll in 1974 by the United States government.

9   "Hundreds of these tropical weather . . . " Key components in the birth of a typhoon explained in Marq de Villiers, *Windswept: The Story of Wind and Weather* (New York: Walker, 2007), 180–182; and Michael Allaby, *Encyclopedia of Weather and Climate* (New York: Facts on File, 2002), 489–493.

10 "Bill Bruder, a twenty-year-old Texan . . . " Bruder described his early years in Austin, Texas, enlistment in the navy, tour of duty at Agana, Guam, and tracking Typhoon Freda in numerous interviews, phone calls, and emails with the author from 2012 through 2015.

12 "[The] birthplace of this . . . " Enewetak Atoll and its role in 1950s nuclear testing described in testimony of Marshall Islands senator Jack Ading before the House Committee on Foreign Affairs Subcommittee on Asia, the Pacific and the Global Environment (Washington, DC, July 25, 2007); additional description of the atoll from National Aeronautics and Space Administration satellite image from January 13, 2013 (see at modis.gsfc.nasa.gov/gallery/individual.phb?db_date=2013-01-26).

12 "It lies in a region . . . " Jack Williams, *The Weather Book* (New York: Vintage Books, 1997), 142.

12 "In the waning days . . . " Genesis of Typhoon Freda described by Bill Bruder, a navy aerographer mate who tracked the typhoon from Naval Air Station Agana, Guam.

13 "The name 'typhoon' . . . " For a detailed history of words used to describe cyclones, see Wayne Neely, *The Great Bahamas Hurricane of 1866* (Bloomington, IN: iUniverse, 2011), chapter 1, "The History behind the Word Hurricane and Other Tropical Cyclone Names."

13 "Cyclones are the biggest . . . " Estimating the number of total global fatalities from cyclones is speculative, but numbering several million. A tally of the twenty-five most deadly cyclones in world history totals at least three million dead, according to Weather Underground, a commercial weather service owned by the Weather Channel and based in San Francisco, California (see the list at wunderground.com/hurricane/deadlyworld.asp).

13 "The practice of naming . . . " de Villiers, *Windswept: The Story of Wind and Weather*, 182–183.

13 "The tendency is to . . . " National Weather Service, "Tropical Cyclone Introduction," 2013 (srh.noaa.gov/jetstream/tropics/tc.htm).

14 "Beyond the death . . . " Neely, *The Great Bahamas Hurricane of 1866*, chapter 1.

14 "On December 17 and 18, 1944 . . . " Barrett Tillman, "William "Bull" Halsey: Legendary World War II Admiral," HistoryNet, 2007 (see at historynet. com/william-bull-halsey-legendary-world-war-ii-admiral.htm).

14 "Freda was five hundred miles west . . . " Bruder interviews.

14 "While Bruder and the other . . . " Author interview with former navy radar technician Keith Clark, November 2012.

17 "Clark's account is consistent . . . " See Saffir-Simpson Hurricane Wind Scale, which rates cyclones from category 1 (minimal) to category 5 (catastrophic), appendix, this book.

18 "It was something formidable . . . " Joseph Conrad, *Typhoon*, 1st Forge ed. (New York: Tim Doherty 1999 [original publication 1903]), 59.

19 "Then there's the sheer size . . . " Ernest Zebrowski Jr., *Perils of a Restless Planet: Scientific Perspectives on Natural Disasters* (Cambridge, UK: Cambridge University Press, 1997), 150.

19 "The size of waves . . . " David W. Wang et al., "Extreme Waves under Hurricane Ivan," *Science* 309 (August 2005): 896.

19 "On October 9 . . . " Bruder interviews.

CHAPTER 3: COUNTDOWN TO CALAMITY

21 "The Thursday storm registered . . . " "Huge Storm Hits—Fierce Wind, Rain," *San Francisco Chronicle*, October 12, 1962, A-1.

22 "The furious winds and seas . . . " "Gale Winds Pound County," *Humboldt Standard* (Eureka, California), October 11, 1962, A-1.

22 "To the north . . . " "New Storm Rolling in Toward Coast," Tacoma *News Tribune* (Tacoma, Washington), October 12, 1962, A-1.

23 "Property damage in Gold Beach . . . " Monetary damages associated with the storm are calculated in 2016 dollars unless otherwise noted. A dollar in 1962 is equal to $7.88 in 2016.

23 "Five deaths in northern . . . " "Huge Storm Hits," *San Francisco Chronicle*, October 12, 1962.

23 "He was the only fatality . . . " Bill Whitman writes about his father in a September 22, 2006, entry to the Clark County History website maintained by the Vancouver *Columbian*, Vancouver, Washington (see at columbian. com/history/columbus).

24 "Meanwhile, the remnants of Typhoon Freda . . . " Portland meteorologist Jack Capell chronicles the development of the Columbus Day storm in his 1962 post-storm report, *The Terrible Tempest of the Twelfth* (Portland: Pioneer Broadcasting Company, 1962).

24 "Ships at sea were . . . " The story of the navy's radar picket ships is told in detail in "The Guardian Class Radar Picket Ships," a 137-page manuscript compiled by the YAGR's Radar Picket Ship Association, which consists of navy officers and enlistees who served on the ships (for more information, visit yagrs.org or contact Lee Doyel at yagrs16@cox.net).

24 "At 5 a.m., Friday . . . " Capell, *Terrible Tempest.*

25 "Aviation forecaster Jim Holcomb . . . " and "Portland weather bureau aviation forecaster . . . " Author interviews with former United States Weather Bureau forecaster Jim Holcomb, July 2015 and October 2016, and

an October 12, 2016, public talk by Holcomb at the Senior Studies Institute of Portland Community College in Portland, Oregon, titled "Forecasting the Columbus Day Storm."

26   "A very deep vigorous . . . " Official United States Weather Bureau weather advisory issued by Portland weather bureau office at 10:10 a.m., October 12, 1962. Viewed as Exhibit A in Columbus Day Storm collection, manuscript 1058, Oregon Historical Research Library, Portland.

26   "The wind warning that followed . . . " Author interview with Holcomb.

26   "East to southeast winds . . . " Official United States Weather Bureau wind warning issued by Portland weather bureau office at 10:40 a.m., October 12, 1962. Viewed as Exhibit B, CDS collection, manuscript 1058, Oregon Historical Society Research Library, Portland.

26   "The prediction of winds . . . " Author interview with Holcomb.

26   "It was not an easy . . . " Capell, *Terrible Tempest.*

27   "I spent three years at sea . . . " Author interview with John Hubbard, deckhand aboard the USS *Tracer* during the Columbus Day Storm, February 2014.

28   "Fishing and other . . . " "The Guardian Class Radar Picket Ships."

28   "Fifty-five foot seas were . . . " Author interview with Carl Coad, assistant navigator aboard the USS *Picket* during the Columbus Day Storm, February 2014.

28   "Any lingering doubt . . . " Capell, *Terrible Tempest.*

28   "The station's wind-damaged anemometer . . . " Wolf Read, a noted Pacific Northwest windstorm historian, has maintained a Columbus Day Storm peak wind map at climate.washington.edu/stormking/October1962.html since September 16, 2001.

29   "I was immediately . . . " Author interview with Chris Percival, US Coast Guard officer stationed at Cape Blanco LORAN station during the storm, April 2015.

29   "By the noon hour . . . " Capell, *Terrible Tempest.*

29   "At Requa Air Force Station . . . " The storm's advance through Northern California and south coastal Oregon is detailed in Franklin, *West Coast Disaster*, 10–20.

30   "The storm that intensified . . . " Weather advisory issued by US Weather Bureau office in Portland at 1 p.m., Friday, October 12.

31   "Even as late as 1 p.m. . . . " Capell, *Terrible Tempest.*

31   "Holcomb relived his most . . . " Holcomb, "Forecasting the Columbus Day Storm."

31   "The National Aeronautics and Space . . . ." The NASA Television Infrared Observation Satellite Program (TIROS) missions are detailed at http://science1nasa.gov/missions/tiros/.

31   "Compare that to present day . . . " Williams, *The Weather Book*, 190–191.

32   "With the development of digital computers . . . " Cliff Mass, *The Weather of the Pacific Northwest* (Seattle: University of Washington Press, 2008), 212–237.

32   "You can't give the weather forecasters . . . " Author interview with Mass, February 2013.

32   "Despite all the advances . . . " Bill Strever, *And Soon I Heard a Roaring Wind: A Natural History of Moving Air* (New York: Little, Brown, 2016), 218–223.

33   "Al Spady, a big, raw-boned . . . " Spady's storm account comes from Betty Chapman Plude, *Columbus Day Storm 1962 Memories* (Independence, OR: self-published, 2011), 71–72. This is a collection of storm memories compiled and copyrighted by Plude to raise money for the Independence Public Library. Plude graciously granted the author permission to draw from these storm memories.

34   "But earlier in the day . . . " Capell, *Terrible Tempest.*

34   "Any barometric pressure reading . . . " Christopher Burt, *Extreme Weather* (New York: W.W. Norton, 2004), 234.

34   "Portland television weather forecaster . . . " Tomlinson, "Predictor of 'Storm of the Century,'" *Oregonian*, June 19, 2009, B-2.

34   "Capell said goodbye . . . " Author interview with Chuck Wiese, a close friend and understudy to Capell, June 2015.

34   "Michael Korte was a . . . " Michael Korte submittal to Vancouver *Columbian* history website, October 2, 2002 (www.columbian.com/history/columbusdaystorm/).

35   "Bob Dernedde remembered . . . " Plude, *Columbus Day Storm 1962 Memories*, 46.

35   "Jennifer White was a . . . " Author interview with Jennifer White, March 2013.

35   "We live submerged . . . " Evangelista Torricelli's scientific discoveries come to life in Gabrielle Walker, *An Ocean of Air: Why the Wind Blows and Other Mysteries of the Atmosphere* (Boston: Houghton, Mifflin, Harcourt, 2007), chapter 1, 7–11. Torricelli is credited with opening the door to the world of wind and meteorology in Jan DeBlieu's *Wind: How the Flow of Air Has Shaped Life, Myth, and the Land* (New York: Houghton Mifflin, 1998), 32.

36   "Other scientists and philosophers . . . ." Walker, *An Ocean of Air*, 14–15.

36   "By the late 1800s . . . " "Brief History of the Barometer" (barometer.ws/history.html).

CHAPTER 4: DEATH COMES TO EUGENE

37   "In the fall of 1962 . . . " Author interview with Sam Grubb in March 2014. Grubb, a University of Oregon student in Eugene in 1962, was witness to one of the most gruesome fatal accidents linked to the Columbus Day Storm.

37   "Just before 4 p.m. at the Eugene airport . . . " Robert E. Lynott and Owen P. Cramer, "Detailed Analysis of the 1962 Columbus Day Windstorm in Oregon and Washington," *Monthly Weather Review* 94, no. 2 (February 1966): 110.

38   "Johnson and Grubb grabbed . . . " Author interview with Grubb.

39   "South of the Amazon Housing Project . . . " "Five Dead, 45 Injured in Eugene Area," Eugene *Register-Guard* (Eugene, Oregon), October 13, 1962, A-1; and "Funeral Scheduled for Storm Victims," Eugene *Register-Guard*, October 14, 1962, A-3.

40   "But the death toll in Eugene . . . " Lucile Payne's first-person account, "Storm Saga Begins: 'It Certainly Was Windy,'" Eugene *Register-Guard*, October 15, 1962, B-1.

41   "The storm was responsible . . . " The Oregon storm fatalities include seventeen reported by the *Oregonian* the day after the storm; nine more reported by the state capital city newspaper, the *Oregon Statesman*, in the six days following the storm; and one Oregon State Hospital patient that doesn't show up on any official reports. The nineteen fatalities in Northern California October 11–13 were reported in the *San Francisco Chronicle* on October 15. The seven storm deaths in British Columbia, Canada, were detailed in the Vancouver *Province* on October 13, and the ten deaths in Washington State were reported in the *Seattle Daily Times* on October 15.

42   "I lost my dad to the storm . . . " Author interview with Daphne Lawrence, February 2013.

42   "I don't think he was . . . " Author interview with Jennifer White, March 2013.

43   "Back then, when doctors . . . " Author interview with Dr. Christopher Wolfe, March 2014.

43   "In addition, more than 42 percent . . . " Anthony Komaroff, "Surgeon General's 1964 Report: Making Smoking History," posted at *Harvard Health Blog*, January 10, 2014 (see at health.harvard.edu).

44   "Trauma from the fatal accident . . . " Author interview with Grubb.

44   "Johnson, a Harvard Law School graduate . . . " Author interview with Art Johnson, April 2014, and copies from Circuit Court of the State of Oregon, Lane County, of amended complaint filed November 30, 1964, and motion of dismissal filed December 10, 1964, and signed by Circuit Judge Edward Levoy the same day.

44   "One bizarre footnote . . . " "Convicted Killer Defiant in Face of a Life Term," KVAL News (Eugene, Oregon), March 25, 2014.

CHAPTER 5: COASTAL CHAOS

47   "The US Coast Guard and . . . " The history of Newport, Oregon, comes alive in Steve Wyatt's *The Bayfront Book* (Waldport, OR: Oldtown Printers, 1999); see page 95, "Guarding the Coast, Yaquina Bay's Life Savers."

47   "At 4 p.m., Coast Guard seaman . . . " Scott McArthur, "Lincoln County Loss May Hit $5 Million," *Capital Journal* (Salem, Oregon), October 15, 1962, section 2, page 13.

47   "The early pioneers . . ." Richard L. Price, *Newport, Oregon, 1866–1936: Portrait of a Coastal Resort* (Newport, OR: Lincoln County Historical Society, 1975), 1.

48   "Early on, the sumptuous . . . " The birth of the oyster industry and construction of the first resort hotel on the Oregon coast is recounted in Wyatt, *The Bayfront Book*, 3–25.

48   "And what brewpub . . . " Wyatt, *The Bayfront Book*, 187–201.

48   "Newport weather on . . . " Tye W. Parzybok, *Weather Extremes of the West* (Missoula, MT: Mountain Press, 2005), 28.

49   "I've lived in Newport since 1941 . . . " Author interview with former Newport fire chief Don Rowley in October 2013.

49   "Don Davis, who arrived in town . . . " Author interview with former
     Newport city manager Don Davis in October 2013.

49   "Newport was punched with wind . . . " Wolf Read, Pacific Northwest
     windstorm historian, compiled a Columbus Day Storm peak wind gust
     map, which is maintained and updated at climate.washington.edu/
     stormking/October1962.html.

50   "Bayfront, which confronted the wind . . . " McArthur, "Lincoln County Loss,"
     Capital Journal, October, 15, 1962, section 2, page 13.

50   "The winds tore off the storefront . . . " Cindy McEntee, *Mo's on the
     Waterfront* (Wilsonville, OR: BestSeller Books, 2004), 46.

50   "The hurricane-force winds . . . " "High Winds Hit Coastal Areas," *Newport
     News* (Newport, Oregon), October 18, 1962, A-1.

50   "You can see why . . . " Author interview with Newport business owner Bill
     Jernigan, October 2013.

51   "The widespread damage . . . " The more than $39 million damage in
     Newport is based on 2016 dollars. National Guard call-out, damage to
     Yaquina View Elementary School, and twenty homes destroyed all included
     in McArthur's "Lincoln County Loss," *Capital Journal*, October 15, 1962,
     section 2, page 13.

51   "The winds pushed a Nye Beach . . . " "High Winds Hit," *Newport News*
     (Newport, Oregon), October 18, 1962, A-1.

52   " . . . Ray Wilkinson was giving" "Depoe Bay News Notes," *Newport News*
     (Newport, Oregon), October 18, 1962, section 2, page 9.

52   "Most of the 150 vessels . . . " Author interview with Newport fisherman Ray
     Hall, October 2013.

53   "Some twenty-five years after the storm . . . " Daphne Lawrence shared with
     the author in February 2013 a two-page undated memoir of the storm that
     her mother, a Newport weather observer for the US Weather Bureau in
     1962, wrote in the late 1970s or early 1980s.

54   "The mill effluent was piped . . . " "Georgia-Pacific Toledo Mill Approximate
     Historical Timeline" (newportoregon.gov).

55   "I've lived here fifty years . . . " Author interview with retired USFS public
     affairs officer Carol Johnson, June 2016.

55   "Rising 3,175 feet from dense . . . " "Mt. Hebo," The Pew Charitable Trust fact
     sheet, September 24, 2013.

56   "In the late 1950s . . . " Eric Schlosser, *Command and Control: Nuclear
     Weapons, the Damascus Accident and the Illusion of Safety* (New York:
     Penguin Books, 2013), 153.

57   "The 160 or so men and women . . . " Author interview with retired Air Force
     Captain David Casteel, March 2014.

57   "The bad thing about Mount Hebo . . . " Author interview with former Air
     Force Airman Second Class Robert Tillmon, March 2014.

57   "Icicles that morphed into lethal . . . " Casteel, "Recollections of the AN/FBS-
     24 Radar at Mt. Hebo Air Force Station," October 12, 2013 (radomes.org).

58   "The double-whammy storm arrived . . . " Mark Cole, stationed at Mount
     Hebo AFS the day of the storm, spoke at a fifty-year storm anniversary
     event at Kane Hall, University of Washington, Seattle, October 11, 2012.

Videos of presentations can be seen at atmos.washington.edu/videos/ Columbus Day/.

59   "A Russian bomber could . . . " Schlosser, *Command and Control*, 178.

61   "Newlyweds Don and Andrea Jenck . . . " The story of the storm and the Jenck dairy farm comes from author interview with Andrea Jenck, October 2013; and "Memories of Destruction," *Tillamook Headlight Herald* (Tillamook, Oregon), October 10, 2012, A-8.

62   "It was the most devastating thing . . . " Ron Zerker, "Memories of Destruction."

63   "The livestock losses spread into the Willamette Valley . . . " Franklin, *West Coast Disaster*, 93–94.

CHAPTER 6: GROUND ZERO

65   "It would take a hurricane . . . " Capell, *Terrible Tempest.*

65   "The US Weather Bureau observer . . . " Weather historian Christopher Burt, "50th Anniversary of the Columbus Day Storm," *WunderBlog*, October 12, 2012.

65   "Exposed ridgetops and hills were . . . " Plude, *Columbus Day Storm 1962 Memories*, 97.

66   "Student Wes Luchau pressed a Graflex . . . " Author interview with Wes Luchau, May 2013.

71   "Ray Coleman was athletic director . . . " Plude, *Columbus Day Storm 1962 Memories*, 35.

71   "Near Eugene, Oregon, Monroe and . . . " Randi Bjornstad, "The Storm of the Century's Destruction, Horror and Wonder Will Never Be Forgotten by Those Who Lived Through It," Eugene *Register-Guard*, October 7, 2012, A-1.

72   "Shoreline High School punter Tom White . . . " Craig Smith, "Punting into This Storm Sent Averages Plummeting," *Seattle Times*, October 12, 2004, D-8.

72   "Perched on the hill was . . . " The tragic and sometimes triumphant history of the Oregon State Hospital is well described in Diane L. Goeres-Gardner, *Inside Oregon State Hospital* (Charleston, SC: History Press, 2013), 41.

73   "They'll put you out at the end . . . " James R. Robblett, as told to *Oregonian* reporter Paul Hause, "Here I Am in an Insane Asylum," *Oregonian*, June 14, 1936, 1.

73   "The northern view from the old . . . " The record population for the hospital was 3,545 in 1958 (Goeres-Gardner, *Inside Oregon State Hospital*, 23).

73   "Presiding over the hospital . . . " Goeres-Gardner, *Inside Oregon State Hospital*, 208–219.

74   "An avid rock climber . . . " Peter Earley, "A Good Friend and Fabulous Advocate Has Died: Dr. Dean Brooks," June 3, 2013), peterearley.com.

74   "He didn't hesitate . . . " Author interview with Brooks's daughter, Dr. Ulista Brooks, May 2013.

74   "Informal by nature . . . " Goeres-Gardner, *Inside Oregon State Hospital*, 211.

74   "Brooks was driving back to the hospital . . . " Author interview with Dr. Dean Brooks, May 2013.

75   "One of the falling oaks . . . " Brooks describes storm fatalities at the hospital in his monthly report to the Oregon Board of Control, November 1, 1962, Oregon State Archives, Salem, Oregon.

75   "The scene Johnson had witnessed . . . " "Storm Damage in Marion County Estimated at $8 Million," *Oregon Statesman*, October 14, 1962, A- 8; "Columbus Day Storm: Worst Disaster Ever," Salem Online History, a project of the Salem Public Library (see at salemhistory.net/natural_history/columbus_day_storm_1962html).

75   "Floyd McKay, a young reporter . . . " Author interview with Floyd McKay, March 2015.

76   "The Oregon state capitol campus . . . " "Columbus Day Storm," Salem Online History.

76   "The popular Republican governor . . . " Plude, *Columbus Day Storm 1962 Memories*, 150.

76   "The governor and his ad hoc . . . " "Portland Radio History," clip no. 7, pdxradio.com.

76   "The governor declared a state . . . " William Hilliard, "Governor Sets Cost in Oregon at $150 Million," *Oregonian*, October 14, 1962, A-1.

77   "Back at Oregon State Hospital . . . ." Goeres-Gardner, *Inside Oregon State Hospital*, 215.

77   "I went ward to ward . . . " Author interview with Dr. Dean Brooks.

78   "Three patients in a medium-security ward . . . " "Patients Try 2-Story Leap, Three Injured," *Oregon Statesman*, October 18, 1962, A-9.

78   "It was necessary to evacuate . . . " Brooks's November 1, 1962, report to the Oregon Board of Control, at Oregon State Archives, Salem, Oregon. Formed by the state legislature in 1913, the board of control provided management oversight of state institutions. Members included the governor, secretary of state, and state treasurer. The board was disbanded in 1969.

79   "I hated everything about it . . . " Author interview with Dr. Dean Brooks.

79   "Some ninety patients and hospital staff . . . " Author interview with former Oregon State Hospital support services director Ray Tipton, April 2013.

CHAPTER 7: A WIND LIKE NO OTHER

83   "Ted Buehner, a weather warning . . . " Gale Flege, "Columbus Day Storm Remains a Fearsome Memory," the *Herald* (Everett, Washington), October 11, 2012, A-1.

83   "Pacific Northwest weather guru . . . " and "The curly-haired professor . . . " Mass, *The Weather of the Pacific Northwest*, 85.

83   "For instance, the Columbus Day Storm . . . " From Columbus Day Storm presentation by Mass at Kane Hall, University of Washington, Seattle, October 11, 2012 (see video at atmos.washington.edu/videos/columbusday).

83   "Storm damage weighted to modern . . . " Estimates of property damage from the storm range from $225 million to $260 million in 1962 dollars or $1.8 billion to $2.1 billion in 2016 dollars, according to Robert E. Lynott and Owen P. Cramer in "Detailed Analysis of the 1962 Columbus Day

Windstorm in Oregon and Washington," *Monthly Weather Review* 94, no. 2 (February 1966), 105. These estimates do not include the value of the roughly 20 percent of windblown timber that could not be salvaged, which, in today's dollars, would be worth at least $1 billion. See *Windthrown Timber Survey in the Pacific Northwest 1962,* Pacific Northwest Region, United States Forest Service (Portland, Oregon, March 1963), 3.

84   "Christopher Burt, a meteorologist . . . " Christopher Burt, "50th Anniversary of the Columbus Day Storm," *WunderBlog,* October 12, 2012.

84   "When Sandy made landfall . . . " Andrew Freedman, "NWS Confirms Sandy Was Not a Hurricane at Landfall," February 13, 2013 (www.climatecentral. org).

84   "They were both extratropical storms . . . " Williams, *The Weather Book,* 140.

85   "Winds topping 100 miles per hour . . . " From Read's Columbus Day Storm peak wind map.

85   "By comparison, the peak gust . . . " Superstorm Sandy wind gusts, storm surge, and barometric pressure readings from Burt, "Hurricane Sandy," November 2, 2012 (blog post at wunderground.com). Columbus Day Storm lowest landfall barometric pressure reading from US Weather Bureau state climatologist Gilbert Sternes, "Summary of the 1962 Columbus Day Storm in Oregon," November 23, 1962 (US Weather Bureau Office, Portland, Oregon), 4.

85   "Even after Superstorm Sandy . . . " Author interview with Mass, September 2014.

86   "Wolf Read, a Seattle native and climatologist . . . " Multiple author interviews with Wolf Read, arguably the foremost scholar and researcher of Pacific Northwest windstorms, beginning in January 2004.

87   "Retired National Weather Service meteorologist . . . " Retired meteorologist George Miller's talk, titled "The Columbus Day Storm October 12, 1962: Has It Happened Before? Will It Happen Again?" Miller was the guest speaker at the Washington State Capital Museum, Olympia, Washington, January 28, 2013.

87   "Read's windstorm curiosity led . . . " Read, "A Climatological Perspective of the 1962 Columbus Day Storm," presented at the Oregon chapter of the American Meteorological Society meeting in Portland, Oregon, October 13, 2012.

88   "In his 2008 Pacific Northwest . . . " Mass, *The Weather of the Pacific Northwest,* 90–94.

89   "For Pacific Northwest tribes . . . " Albert Reagan and L. V. W. Walters, "Tales from the Hoh and Quileute," *Journal of American Folklore* 46, no. 182 (1933): 297–346.

89   "The force of Southerly winds . . . " John Meares, *Voyages Made in the Years 1788 and 1789 from China to the Northwest Coast of America* (New York: Da Capo Press, 1967 [original publication 1790]), 253.

89   "The white explorers and pioneers . . . " George R. Miller, *Lewis and Clark's Northwest Journey: Weather Disagreeable* (Portland: Frank Amato Publications, 2004), 21–22.

90   "William Kitson, the Hudson's Bay factor . . . " Lois Brandt Phillips, "Cold, Snow More Painful in Early Days," *News Tribune* Sunday Ledger magazine (Tacoma, Washington), January 12, 1969, page 3.

90   "The deadliest and perhaps . . . " David Wilma, "Several Dozen Fishermen Drown Off the Mouth of the Columbia River on May 4, 1880," HistoryLink. org (Essay 7932, October 14, 2006).

90   "The west side of the Olympic Peninsula . . . " Read, "The Olympic Blowdown of January 29, 1921," November 23, 2013, (climate.washington.edu/ stormking/January1921.html).

91   "Another major windstorm struck . . . " Mass, *The Weather of the Pacific Northwest*, 85.

91   "The same holds true with a review . . . "Major Southern California Windstorms (1858–November 2013)," 2014 Natural Hazards Mitigation Plan, City of Newport Beach, California, California, Section 10, Table 10-4, pages 7–12.

92   "In the case of the Columbus Day Storm . . . " Capell, *Terrible Tempest.*

92   "The low-pressure reading . . . " Mass, *The Weather of the Pacific Northwest*, 87; and appendix, this volume.

92   "The storm plunged . . . " Oregon meteorologist Steve Pierce, "The Columbus Day Storm of 1962," *Oregon Chapter of the American Meteorological Society Newsletter* (Portland), October 2, 2012.

92   "The storm center was close . . . " Some of the best descriptions of what made the Columbus Day Storm so strong are found in George R. Miller's *Pacific Northwest Weather* (Portland: Frank Amato Publications, 2002), 67–74.

92   "If the winds had come . . . " Mark Floyd, "Fifty Years Later: Legacy of Columbus Day Storm Still Stands," featuring Kathy Dello, deputy director of Oregon Climate Science at Oregon State University, OSU News and Research (Corvallis), October 2, 2012.

92   "Adding to the storm's might . . . " Lynott and Cramer, *Detailed Analysis of the 1962 Columbus Day Windstorm*, 116.

93   "The surface winds were enhanced . . . " Author interview with Wolf Read, March 2015.

93   "And the sheer size . . . " Ellis Lucia, *The Big Blow* (Portland: New-Times Publishing, 1963), 65; and undated storm report by US Weather Bureau state climatologist Earl L. Phillips, "Columbus Day Storm in Washington, October 12, 1962," page 1 (retrieved from Washington State Library, Olympia, January 2013).

93   "The Columbus Day Storm arrived . . . " George H. Taylor and Raymond H. Hatton, *The Oregon Weather Book: A State of Extremes* (Corvallis: Oregon State University Press, 1999), 139.

93   "Fifty years after the . . . " Pierce, Oregon, chapter of *American Meteorological Society Newsletter*, October 2, 2012.

93   "Without a leveling off . . . " *The Third Oregon Climate Assessment Report*, Oregon Climate Change Research Institute (Corvallis: Oregon State University), January 2017.

94   "Neither climate model projections . . . " Russell S. Vose et al., "Monitoring
     and Understanding Changes in Extremes: Extratropical Storms, Winds and
     Waves," *Bulletin of the American Meteorological Society* (March 2014), 3.

94   "Climate scientists Christian Seiler and F. W. Zwiers . . . " "How Will Climate
     Change Affect Explosive Cyclones in the Extratropics of the Northern
     Hemisphere?" *Climate Dynamics*, August 12, 2015.

94   "There's no reason to expect Northwest windstorms . . . " Cliff Mass,
     "Extreme Weather Trends over the Pacific Northwest," paper presented at
     the Northwest Climate Conference, University of Washington, Seattle,
     September 9, 2014.

95   "There's no evidence that extratropical cyclones . . . " Author interview with
     Oregon Climate Change Research Institute associate director Kathie Dello,
     November 2015.

CHAPTER 8: FALLEN FORESTS

97   "Three of the five tallest . . . " Stephen F. Arno and Ramona P. Hammerly,
     Northwest Trees: Identifying and Understanding the Region's Native Trees
     (Seattle: Mountaineers Books, 2007), 57.

97   "Forestlands represent roughly . . . " Five Year Forest Inventory and Analysis
     Report for Washington (report for 2002–2006, Research Bulletin GTR-800,
     April 2010), and Five Year Forest Inventory and Analysis Report for Oregon
     (report for 2001–2005, Research Bulletin GTR-765, November 2008),
     prepared by US Forest Service Pacific Northwest Station, Portland, Oregon.

97   "About half of the . . . " *Windthrown Timber Survey in the Pacific Northwest
     1962,* Pacific Northwest Region (United States Forest Service, Portland,
     Oregon, March 1963), 3.

97   "An estimated fifteen billion board feet . . . " *Western Conservation Journal*
     20, no. 4 (1963): 24.

98   "The Columbus Day Storm windfall . . . " The estimate of enough timber to
     frame nearly one million average-sized homes is based on an Idaho Wood
     Products Commission fact sheet available at idahoforests.org/woodhouse.

98   "It's a volume more than twice . . . " Total timber harvest in Washington and
     Oregon in 2014 was 7.33 billion board feet, according to annual reports
     issued by Washington State Department of Natural Resources, Olympia,
     August 2015, and Oregon State Department of Forestry, Salem, July 2015.

98   "It was more than three times . . . " Timber damaged or destroyed by the May
     1980 eruption of Mount St. Helens is estimated at 4.7 billion board feet.
     Rob Carson, *Mount St. Helens: The Eruption and Recovery of a Volcano*,
     35th anniv. ed. (Seattle/Tacoma: Sasquatch Books/News Tribune, 2015
     [(original publication 1990]), 53.

98   " . . . a greater loss than what occurred . . . " *Western Conservation Journal* 20,
     no. 4: 24.

98   "By the time the Columbus . . . " Charles L. Bolsinger and Karen L. Waddell,
     Area of Old-Growth Forests in California, Oregon and Washington, US
     Forest Service, Research Bulletin 197 (Portland: Pacific Northwest Research
     Station), December 1993, 3.

98   "The Sitka spruce stands . . ." Wyatt, *The Bayfront Book,* 137–150.

98   "An old-growth inventory . . ." *2002–2006, Five Year Forest Inventory and Analysis Report* (Washington).

98   "Douglas-fir *(Pseudotsuga menziesii)* . . ." Invaluable in defining the botanical, cultural, and economic value of the conifer trees that took the biggest beating in the Columbus Day Storm were Donald C. Peattie, *A Natural History of Western Trees* (Boston: Houghton Mifflin, 1950); Robert Van Pelt, *Forest Giants of the Pacific Coast* (Seattle: University of Washington Press, 2001); David Suzuki and Wayne Grady, *Tree: A Life Story* (Vancouver, BC: Greystone Books, 2004), and the aforementioned Arno and Hammerly, *Northwest Trees.*

100 "Anderson was driving . . ." Les Joslin, "Old Smokeys Remember the Big Blow," *OldSmokeys Newsletter* (Pacific Northwest Forest Service Association, Portland, fall 2012). Newsletters are archived at www.oldsmokeys.org.

100 "It didn't take long Friday night . . ." Author interview with former state forestry manager Glen Hawley, January 2013.

101 "Public and private timberland owners . . ." John Clark Hunt, "The Great Columbus Day Blowdown," *American Forests* (January 1963), 15, 52.

102 "It was before the national and state . . ." Author interview with former state Department of Natural Resources forester Kenhelm Russell, January 2013.

102 "There was a sense of urgency . . ." *Western Conservation Journal* 20, no. 4 (1963): 42.

103 "Neil Phillips, a US Forest Service . . ." Plude, *Columbus Day Storm 1962 Memories,* 162.

103 "The powers that be approved . . ." Joslin, "Old Smokeys Remember."

103 "The first timber sale . . ." Author interview with retired US Forest Service planner Ken McCall, February 2017.

104 "Timber blown down or . . ." Author interview with Hawley.

104 "A lot of times you were . . ." Author interview with retired DNR forester Roy Friis, January 2013.

104 "The Columbus Day Storm touched off . . ." *Western Conservation Journal* 20, no. 4: 56.

104 "On the Siuslaw National . . ." Author interview with McCall.

104 "Logging roads boost the likelihood . . ." John Daniel, *The Far Corner: Northwestern Views on Land, Life and Literature* (Berkeley: Counterpoint Press, 2009), 75.

106 "The storm struck just as Japan's . . ." Jean M. Daniels, *The Rise and Fall of the Pacific Northwest Log Export Market,* United States Forest Service, General Technical Report 624 (Pacific Northwest Research Station, Portland, Oregon, February 2005), 5–7; Daniel Jack Chasan, *The Water Link: A History of Puget Sound as a Resource* (Seattle: Washington Sea Grant Program, University of Washington, 1981), 118–119.

106 "In export, hemlock was the preferred . . . " Joni Sensel, *Tradition through the Trees: Weyerhaeuser's First 100 Years* (Seattle: Documentary Book Publishers, 1999), 127.

106 "Log exports had just begun . . . " Daniels, *The Rise and Fall*, 5, figure 3.

107 "I remember it like it was yesterday . . . " Author interview with forester Bob Dick, July 2013.

107 "Out of the storm grew . . . " Sensel, *Tradition through the Trees*, 127–136.

107 "High-yield forestry had its detractors . . . " John G. Mitchell, "Best of the S.O.B.s," *Audubon* magazine (September 1974): 51.

108 "The fallen forests . . . " Fresh looks at forest ecology are found in Peter Wohlleben, *The Hidden Life of Trees* (Vancouver, BC: Greystone Books, 2015); and Suzuki and Grady, *Tree: A Life Story.*

CHAPTER 9: THE WIND AND WINE

109 "Those trees bent, twisted, and . . . " George W. Moore, *Geology of Vineyards in the Willamette Valley* (Corvallis: Oregon State University, Department of Geosciences), 4. Updated September 16, 2002.

109 "A tally of the storm damage . . . " "Orchard Loss High: Some May Be Saved," *Newberg Graphic* (Newberg, Oregon), October 25, 1962, 2-1.

110 "The grim losses played out . . . " Floyd McKay, "Valley Orchard Damage May Hit $8 Million" [in 1962 dollars], *Oregon Statesman* (Salem, Oregon), October 19, 1962, A-1.

110 "Jay Greer, owner of a forty-acre . . . " Author interview with Salem area farmer Jay Greer, May 2013.

110 "Newberg orchardist Bradley Smith . . . " "NW Farms Make Big Storm Recovery," *Oregonian*, October 13, 1963, page A-1.

110 "Farmers were faced with tough . . . " McKay, "Valley Orchard Damage," *Oregon Statesman.*

110 "The Oregon walnut industry . . . " Nicole Montesano, *Hull of a Good Story* (McMinnville, OR: Oregon Wine Press, 2012).

110 "The state's prune industry . . . " "Furious Typhoon Gales Strike Area," *Newberg Graphic*, October 18, 1962, A-1; and "Oregon Agripedia," an Oregon Department of Agriculture annual report (2015), 4 (see the full report online at oregon.gov/ODA/shared/Documents/Publications/.../Agripedia.pdf).

110 "Only filberts, more commonly known . . . " Crop statistics found in "Oregon Agripedia" (2015), 2–6.

111 "There is another crop and industry . . . Christian Miller, *The Economic Impact of the Wine and Grape Industries on the Oregon Economy* (Berkeley: Full Glass Research, 2015); and "Oregon Agripedia" (2013).

111 "While orchardists in the north Willamette . . . " Much has been written about the iconic Oregon wine pioneers of the north Willamette Valley, Chuck Coury and David Lett. The author relied largely on two reports by aficionado William "Rusty" Gaffney, from his online newsletter, PinotFile, including "Oregon Relishes Burgundian Influence" (vol. 6, no. 36, August 13, 2007); and "Oregon Pinot Noir: Who Planted First?" (vol. 8, no. 39, June 10, 2011). All of the PinotFile newsletters are archived at princeofpinot.com.

Lett's life is summarized by Eric Asimov, "David Lett, Oregon Wine Pioneer, Dies at 69," *New York Times*, October 14, 2008, B-14. It's important to note that the original 1840s settlers in the Willamette Valley planted grapes and made wine, and the first modern-day Pinot Noir grape grower and winemaker was Richard Sommer of Hillcrest Vineyard in Roseburg in the Umpqua Valley in 1961.

112 "Another Oregon wine-pioneering couple . . . " The Oregon Wine History Project at Linfield College in McMinnville, Oregon, is a treasure trove of oral histories and photographs from the early days of the Oregon wine industry. See the Dick Erath interview transcript, July 8, 2010, at digitalcommons@linfield.edu.

112 "The wine pioneers intent on growing . . . " Author interview with Susan Sokol Blosser, September 2015; and Susan Sokol Blosser, *At Home in the Vineyard: Cultivating a Winery, an Industry, and a Life* (Berkeley: University of California Press, 2006), 8–9; and Susan Sokol Blosser interview transcript, May 22, 2012, digitalcommons@linfield.edu.

112 "One of those farmers, who predated . . . " "Jim Maresh: Old Roots Run Deep," Avalon Wine and Northwest Wine (Portland, northwest-wine.com/Maresh-Vineyard-Family.php); Katherine Cole, "A Dundee Hills Name to Know This Wine Month," OregonLive.com (Portland), May 5, 2015.

113 "The conversion of storm-battered . . . " Author interview with Elk Cove winery owner Adam Campbell, September 2015; and Vivian Perry and John Vincent, *Winemakers of the Willamette Valley: Pioneering Vintners from Oregon's Wine Country* (Charleston, SC: American Palate/History Press, 2013), 73–76.

CHAPTER 10: BRIDGETOWN UNDER SIEGE

117 "Busy but not in a hurry . . . " Describing the evolution of Portland as a Pacific Northwest metropolitan city is the forte of Portland State University urban studies professor Carl Abbott in *Greater Portland: Urban Life and Landscape in the Pacific Northwest* (Philadelphia, PA: University of Pennsylvania Press, 2001); and *Portland in Three Centuries: The Place and the People* (Corvallis: Oregon State University Press, 2011).

117 "The Morrison Street Bridge . . . " Barner Blalock, *Portland's Lost Waterfront* (Charleston, SC: History Press, 2012), 91–96.

117 "At the peak of the storm . . . " James Lattie, "Giant Mop-Up Project Begins in Gale's Wake," *Oregonian*, October 14, 1962, A-1.

117 "But that early fall storm . . . " A sunny day is defined as 30 percent or less cloud cover (see currentresults.com/weather/Oregon/annual-days-of-sunshine.php).

118 "The airwatch pilot Friday night . . . " "Flier Takes Bumpy Ride," *Oregonian*, October 14, 1962, A-46.

118 "Some 175 planes were tossed around . . . " Leverett Richards, "175 Planes Casualties of Typhoon in One of Flying's Darkest Days," *Oregonian*, October 14, 1962, A-39.

118 "Two dozen airplane hangars . . . " Lucia, *The Big Blow* (Portland: New-Times Publishing, 1963), 64.

118 "Every time Wikander . . . " "Flier Takes Bumpy Ride," *Oregonian*.

119 "Wikander, a competitive, accomplished aviator . . . ." Personal details about Wikander from author interview with her friend and former coworker, Wayne Nutsch, April and May 2014. Physical description of Wikander from historic photo in Ninety-Nines, Inc., newsletter, *New Horizons* (November-December 2008): 31.

119 "As Wikander fought to get . . . " Author interview with Capell protégé, Chuck Wiese, June 2015.

119 "Almost out of fuel . . . " "Flier Takes Bumpy Ride," *Oregonian*.

119 "Airplane carnage was widespread . . . " Richards, "175 Planes Casualties," *Oregonian*.

121 "As the winds picked up pace . . . " Francis Murnane report submitted December 12, 1962, to Oregon Historical Society, Columbus Day Storm Collection, manuscripts 1038.

122 "Damage to yacht clubs and houseboats . . . " "Boats, Large and Small, Towed Back to Berths," *Oregonian*, October 14, 1962, A-31.

122 "Two-year-old Michael Gensel . . . " "Giant Mop-Up Project," *Oregonian*, October 14, 1962, A-1.

122 "Leo J. Buyseries, a prominent farmer . . . " "Hot Wire Fatal, Toll in Valley Grows," *Oregon Statesman*, October 15, 1962, A-1.

122 "Twenty years before the storm . . . " "Japanese Wartime Incarceration in Oregon," Oregon Encyclopedia of History and Culture, a joint project of Portland State University and the Oregon Historical Society (browse topics alphabetically at oregonencyclopedia.org/).

123 "A fourth storm fatality in Portland . . . " "Giant Mop-Up Project," *Oregonian*.

123 "Storm-related injuries sent . . . " Franklin, *West Coast Disaster*, 114.

123 "The storm chewed up suburbs . . . " "Flight Made by Building," *Oregonian*, October 14, 1962, A-43.

123 "The winds held Albert and Hazel . . . " William Swing, "Couple Rides Out Storm as House Comes Down," *Oregonian*, October 14, 1962, A-39.

124 "City parks and the woods . . . " Tree loss in Portland was estimated at ten thousand in *Summary of the 1962 Columbus Day Storm in Oregon*, Gilbert Sternes, US Weather Bureau Office, Portland (November 23, 1962), 7; and sixteen thousand in "Columbus Day Storm, The Benchmark for PacNW Windstorms," Tyree Wilde, National Weather Service, talk presented at the American Meteorological Association meeting October 13, 2012, Portland, Oregon.

124 "Ted Buehner, a weather warning meteorologist . . . " Author interview with National Weather Service meteorologist Ted Buehner, March 2013.

124 "In 5,100-acre Forest Park, . . . " Information on Forest Park and Columbus Day Storm damage available at www.portlandoregon.gov/parks.

125 "Storm damage proved to be a blessing . . . " Visit the Pittock Mansion at 3229 NW Pittock Drive, Portland, Oregon, to learn about the mansion's history and how the Columbus Day Storm "saved" the mansion from demolition.

125 "Murnane, the son of immigrant Irish parents . . . " Author interview with Portland author and union labor activist Michael Munk, September 2014;

and Munk's unpublished profile, "Francis J. Murnane: Oregon's Forgotten 'Rabble-Rousing, Art-Loving Longshoreman,'" June 2012.

126 "As the winds subsided . . . " Murnane's account of storm submitted to Oregon Historical Society, December 12, 1962.

CHAPTER 11: LIFE TURNS ON A DIME

129 "The hundreds of houseboats . . . " "Boats, Large and Small, Towed Back to Berths," *Oregonian*, October 14, 1962, A-31.

129 "The storm claimed three more . . . " Bill Sieverling, "Deaths, Damage Remain in Wake of Gale's Path," *Columbian* (Vancouver, Washington), October 15, 1962, A-1.

129 "At the top of Mount Pleasant . . . ." "Injured Man Receives Aid," weekend edition of Longview *Daily News* (Washington), October 13–14, 1962, A-1.

130 "Fifteen miles southeast of Olympia . . . " Author interview with Susan Archibald, July 2015.

130 "State Patrol trooper Emmett Smith . . . " Author interview with former Washington State Patrol trooper Emmett Smith, October 1987.

131 "In the summer of 2015 . . . " Author interview with Susan Archibald, July 2015.

131 "Newspaper accounts of the Archibald parents . . . " "Yelm Parents Are Killed as Tree Topples on Car," Sunday *Olympian* (Olympia, Washington), October 14, 1962, A-1.

132 "Carlene Pohl, a nurse's aide . . . " Author interview with Carlene Pohl, September 2013.

133 "The storm came to life for me . . . " Author's memories of the Columbus Day Storm.

134 "I was cooking oven fried chicken . . . " Author interview with Charlotte Kilde, March 2013.

134 "After the winds finally died down . . . " Author's memories of the Columbus Day Storm.

CHAPTER 12: LIONS IN THE WIND

139 "The storm strengthened as Charley Brammer . . . " Author interviews with Charley Brammer and Ray Brammer, December 2012.

142 "Speaking to a TNT reporter . . . " "Lions Attack Woman, Boy," *News Tribune* (Tacoma, Washington), October 13, 1962, A-1.

143 "Zoo officials said they found no . . . " Kris Sherman, public relations coordinator at Point Defiance Zoo and Aquarium in Tacoma, Washington, responding to author's June 2013 public information request about lion populations and policies at the zoo.

143 "The twin lionesses that escaped . . . ." Sean Robinson, "He'll Never Forget the Columbus Day Storm," *News Tribune* (Tacoma, Washington), October 12, 2002, A-1.

143 "Two months after the mauling . . . " Charles and Ray Brammer complaint, Pierce County Superior Court Case No. 154074, filed December 18, 1962. McCallister response filed January 4, 1963. The dollar figures in the lawsuit and settlement are in 1962 dollars.

144 "This much is clear . . . " *A Life Sentence: The Sad and Dangerous Realities of Exotic Animals in Private Hands in the U.S.* (Sacramento, CA: American Protection Institute, 2006; see at bornfreeusa.org/a3b1_invest.php).

144 "The plaintiff was painfully injured . . . " Helen V. Sullivan complaint, Pierce County Superior Court Case No. 154508, filed January 18, 1963. McCallister response filed February 4, 1963.

145 "At some point during the legal fight . . . " Author interview with Ray Brammer, December 2012.

145 "They decided to forgo the trial . . . " Judgment on behalf of Charles and Ray Brammer, Pierce County Superior Court Judge John D. Cochran, signed September 17, 1963.

145 "In the years that followed . . . " Author interview with Charley Brammer, December 1962.

145 "His dad isn't so sure . . . " Author interview with Ray Brammer, December 1962.

145 "I don't think about it . . . " Author interview with Charley Brammer, December 1962.

145 "We were married a year . . . " Author interview with Debbie Brammer, December 1962.

146 "Not far from the Point Defiance . . . " Terry Terrian, "A Wild Ride on a Fierce Black Night," Tacoma *News Tribune* Sunday Ledger, October 8, 1967, A-6.

146 "Dubbed the "City of Destiny . . . " Murray Morgan, *Puget's Sound: A Narrative of Early Tacoma and the Southern Sound* (Seattle: University of Washington Press, 1979), 332.

146 "The shipmaster's whistle blast . . . " Terrian, "A Wild Ride," Tacoma *News Tribune* Sunday Ledger.

CHAPTER 13: IT HAPPENED AT THE FAIR (BUON GUSTO)

149 "Duff Andrews, twenty-two and a dental school student . . . " Author interview with Seattle World's Fair employee Duff Andrews, April 2014.

149 "We were very scared at first . . . " "Stuck in Needle Elevator, Young Men Play Gin Rummy," *Seattle Daily Times*, October 13, 1962, A-1.

149 "Andrews recalled the elevator bobbed . . . " Author interview with Andrews, April 2014.

151 "The doubters initially included . . . " Alan J. Stein, "Century 21—The Seattle's World Fair, Part 1," HistoryLink.org (Essay 2290, April 19, 2000), 3.

151 "Even Seattle's World Fair colors were . . . " "Space Needle History: Sketchy Beginnings" (spaceneedle.com/history/).

151 "Columbus Day festivities kicked off . . . " Murray Morgan, *Century 21: The Story of the Seattle World's Fair,* 50th anniv. ed. (Bellingham, WA: Chuckanut Editions, 2012 [University of Washington Press, 1963]), 15.

151 "Albert Rosellini, the first Italian-American . . . ." Paula Becker, Alan Stein, and the HistoryLink staff, *The Future Remembered: The 1962 World's Fair and Its Legacy* (Seattle: Seattle Center Foundation), 190.

152 "A keen supporter when the fair was . . . " Paula Becker, Alan Stein, and the HistoryLink staff, *The Future Remembered,* 170–171.

152 "On Tuesday the week of the storm . . . " "Rosellini Asks Car Seat Belts," *Oregonian*, October 12, 1962, A-15.

152 "The son of an Italian immigrant . . . " Rick Anderson, "The Fixer," November 19, 2013 (www.seattleweekly.com/home/949636-129/inside-al-rosellinis-fbi-file).

152 "Frank Colacurcio, one of Seattle's most . . . " Rick Anderson, *Seattle Vice: Strippers, Prostitution, Dirty Money and Crooked Cops in the Emerald City* (Seattle: Sasquatch Books, 2010), 14, 65–66.

153 "Many called Seattle home . . . Matthew Kingle, *Emerald City: An Environmental History of Seattle* (New Haven, CT: Yale University Press, 2007), 101–102.

153 "Chief Seattle, whose people were banned . . . " Jennifer Ott, "Seattle Board of Trustees," History Link.org (Essay 10979, December 7, 2014), 1.

153 "By 6 p.m., wind gusts strong enough . . . " Becker, Stein, et al., *The Future Remembered*, 190.

153 "The Plaza of the States was no longer . . . " "Security Men Urge All to Leave," *Seattle Post-Intelligencer*, October 13, 1962, A- 3.

153 "All over the fairgrounds . . . " Becker, Stein, et al., *The Future Remembered*, 192.

154 "Jack Gahagan, forty-three, a worker in Gayway . . . " "Storm Damage Slight at Fair," *Seattle Daily Times*, October 13, 1962, A-1.

154 "The Chicago fair was approved by Congress . . . " "The Big Wheel," Hyde Park Historical Society newsletter (Spring 2000) (hydeparkhistory.org/newsletter.html).

154 "The novelty of being stuck . . . " Eric Sorenson, "Columbus Day 1962: Memories of Storm That Roared Still Vivid," *Seattle Times*, October 6, 2002, A-1.

154 "Among the nighttime fair attendees . . . " Author interview with Seattle radio personality Jim French, April 2014.

154 "Maybe French had heard . . . " David Laskin, *Rains All the Time* (Seattle: Sasquatch Books, 1997), 10.

155 "I was trying to keep them . . . " "Ferocious Storm Surprised the Pacific Northwest on Columbus Day 1962," October 12, 2002 (usatoday30.usatoday.com/weather/news/2002).

155 "The winds continued to stiffen . . . " "Seattle Trivia, Above All Trivia" (spaceneedle.com/fun-facts/).

156 "The Space Needle grew from a doodle . . . " Knute Berger, *Space Needle: The Spirit of Seattle* (Seattle: Documentary Media, 2012), 30.

156 "With its tripod legs and . . . " (spaceneedle.com/fun-facts/).

156 "The night of the windstorm . . . " Becker, Stein, et al., *The Future Remembered*, 190.

156 "French and his wife walked . . . " Author interview with French.

157 "These were heady times at the Boeing Company . . . ." "727 Commercial Transport: Historical Snapshot" (boeing.com/history/products/727.page).

157 "French described the harrowing . . . " Author interview with French.

157 "The rushing wind turned the needle's . . . " Franklin, *West Coast Disaster*, 149–150.

157 "A few hungry Space Needle refugees . . . " Author interview with Ralph
     Munro, April 2014.

158 "The maritime storm scene in Seattle . . . " "Winds Whip Waters Here; Boats,
     Ferry Runs Halted," *Seattle Post-Intelligencer*, October 13, 1962, A-4.

158 "The search and rescue . . . " Ken Robertson, "Seamen Claim Storm's Waves
     Highest Yet," *Bellingham Herald* (Bellingham, Washington), October 14,
     1962, A-1.

158 "Ron Newell, a nineteen-year-old . . . " Author interview with Ron Newell,
     March 2015.

159 "Just as state officials were . . . " " "Security Men Urge," *Seattle Post-
     Intelligencer*, October 13, 1962.

159 "The other exception to the early fair closure . . . " "Storm Damage Slight,"
     Seattle Daily Times, October 13, 1962, A-1.

160 "The fair finale on Oct. 21 . . . " Becker, Stein, et al., *The Future Remembered*,
     195.

160 "It was a fitting venue . . . " Morgan, *Century 21*, 43.

160 "The powerful and persuasive Magnuson . . . " Becker, Stein, et al., *The Future
     Remembered*, 195.

160 "What Kennedy and Johnson had . . . " Schlosser, *Command and Control*,
     288–297.

161 "When I visited . . . " Kristin Kendle, "Top Ten Tallest Buildings in Seattle,"
     February 12, 2016, (seattle.about.com).

161 "The Space Needle hosts . . . " (spaceneedle.com/fun-facts/).

161 "There is a tangle of industry . . . " Kathy Weiser, "Washington State Legends:
     Harbor Island Largest Artificial Island in the U.S.," December 2015
     (legendsofamerica.com).

CHAPTER 14: TERROR IN STANLEY PARK

163 "The park's popularity is the stuff of legend . . . " Eric Nicol, *Vancouver*
     (Garden City, NY: Doubleday, 1978), 78.

164 "Stanley Park was a scene of terror . . . " Ormond Turner, "Terror among the
     Trees," *Province* (Vancouver, BC), October 13, 1962, A-1.

165 "I took one look and knew . . . " "Man Still Hears Sounds of Death,"
     Vancouver Sun (Vancouver, BC), October 13, 1962, A-3.

165 "Joseph Plag was northbound on the causeway . . . " "Man Still Hears Sounds
     of Death," *Vancouver Sun*.

165 "Photographer Ray Allan of the *Vancouver Sun* . . . " "Causeway a Nightmare,"
     Vancouver Sun, October 13, 1962, A-2.

165 "The storm damaged or destroyed . . . " Author interview with forest
     climatologist Wolf Read, March 2015.

166 " . . . natives who showered British naval explorer . . . ." Nicol, *Vancouver Sun*,
     13.

166 "In the aftermath of the storm . . . " Pat Carney, "Park Will Log, Oops, Clean
     Up Downed Timber," *Province* (Vancouver, BC), October 19, 1962, A-1.

166 "Call it what Lefeaux would . . . " Nicol, *Vancouver Sun*, 233.

166 "The seven other storm-related fatalities . . . " The eight total storm fatalities in the greater Vancouver area is a combined tally from page 1 stories in the *Vancouver Sun* and the *Province*, October 13, 1962.

167 "Others experienced close calls . . . " "Gale Force Winds Worst in History," *Vancouver Sun*, October 13, 1962, A-1.

167 "Louis Kadla spent some ninety minutes . . . " "B.C. Fisherman Swims Ashore When Boat Sank," *Seattle Daily Times*, October 13, 1962, A-6.

168 "The storm knocked out power . . . " "Blackout Worst in B.C.'s History," *Vancouver Sun*, October 13, 1962, A-1.

168 "On a drizzly, windless Sunday . . . " Author tour of Stanley Park and its windstorm history with forest climatologist Wolf Read, March 2015.

170 " . . . a cityscape filled today with . . . " SkyscraperPage.com database search of Vancouver, BC.

170 " . . . in a city where 25 percent . . . " John Punter, *The Vancouver Achievement* (Vancouver: University of British Columbia Press, 2003).

170 "In 2014, Vancouver ranked ninth . . . " Amy Judd, "Vancouver Ranked 9th in the World for Highest Number of Skyscrapers," July 23, 2014 (globalnews. ca/news/1469699).

170 " . . . a rather remarkable ascension . . . " Nicol, *Vancouver Sun*, 60–72, 142.

170 " . . . hoisting my glass to the memory . . . " Nicol, *Vancouver Sun*, 211.

170 "At. 1 a.m., Saturday, Capell and . . . " "Portland Radio History," pdxradio. com.

171 "Capell drove home to his wife . . . " October 18, 1962, letter written by Capell's mother, Mabel Capell, to Anton and Laimi Wagner, the parents of Capell's wife, Sylvia.

CHAPTER 15: A STORMY AFTERMATH

173 "After shredding Vancouver, British Columbia . . . " Author interview with Wolf Read, March 2015.

173 "It marked the third straight day . . . " Burt, "50th Anniversary of the Columbus Day Storm," *WunderBlog*.

173 "By comparison, rainfall in Oregon and . . . " Fred W. Decker, Owen P. Cramer, and Byron P. Harper, "The Columbus Day 'Big Blow' in Oregon," *Weatherwise*, December 1962, 244.

173 "The pulverizing Bay Area rains . . . " Jack Craig, "Landslides and Flood— 'Disaster' in Oakland," *San Francisco Chronicle*, October 14, 1962, A-1.

173 "The first mudslide fatality happened . . . " *San Francisco Chronicle*, October 14, 1962, 1, 27.

174 "Bill McCarthy and his . . . " *San Francisco Chronicle*, October 14, 1962, 27.

174 "A mother and daughter were buried . . . " "Two in Car Buried in Sea of Mud," *San Francisco Chronicle*, October 14, 1962, A-1.

174 "Tragedy was not over yet . . . " "Survey of West Coast Storm Loss," *San Francisco Chronicle*, October 15, 1962, A-10.

174 "On Sunday, two-year-old . . . " "The Big Mud Cleanup—Damage in the Millions," *San Francisco Chronicle*, October 15, 1962, A-1, A-10.

174 "And San Francisco physician . . . "Storm Crash Injuries Fatal to S.F. Doctor," *San Francisco Chronicle*, October 15, 1962, A-13.

174 "It can't hurt my batting eye . . . " Art Rosenbaum, "Yanks, Giants Seek Dry Ground," *San Francisco Chronicle*, October 14, 1962, Sporting Green (sports section), 47.

175 "Taking advantage of the storm delays . . . " William Ryczek, "Game 7 of the 1962 World Series," National Pastime Museum, February 11, 2015 (thenationalpastimemuseum.com).

175 "To the north, the college football . . . " Don Fair, "Stadium Damaged by Winds," *Oregonian*, October 13, 1962, 3-1.

175 "Roofing and other debris . . . " "Friday Night's Violent Winds Wreck Multnomah Stadium," *Register-Guard* (Eugene, Oregon), October 14, 1962, B-1.

176 "After the game, both teams . . . " Bob Robinson, "Cold Quiets Husky Cheer," *Oregonian*, October 14, 1962, 3-1.

176 "Baker stayed in Portland . . . " Author interview with Baker, May 2013.

177 " . . . crippled by widespread power outages . . . " Stuart Tomlinson, "Columbus Day Storm Still Howls," *Oregonian*, October 7, 2012, A-1; author interview with Wolf Read, March 2015.

177 "Retired Portland General Electric . . . " R. J. Brown, a retired PGE lineman, was interviewed in a video presented by KPTV-12 meteorologist Brian McMillan at a public symposium hosted by the Oregon chapter of the American Meteorological Society at the Oregon Museum of Science and Industry, Portland, October 13, 2012.

177 "Longview, Washington, had some three hundred miles . . . " Lucia, *The Big Blow*, 35–36.

178 "Cathy Schoenborn McLean was a . . . " Plude, *Columbus Day Storm 1962 Memories*, 139–140.

178 "Five days after the storm . . . " "To Our Customers," *Corvallis Gazette-Times*, October 17, 1962, A-3.

178 "Some forty-five thousand Oregon families . . . " "45,000   Remain without Power," *Corvallis Gazette-Times*, October 18, 1962, A-1.

178 "Oregon State University forestry students . . . " Plude, *Columbus Day Storm 1962 Memories*, 164.

179 "For a time the sound . . . " Plude, *Columbus Day Storm 1962 Memories*, 70.

179 "McCulloch Corporation shipped 1,100 chain saws . . . " *Western Conservation Journal* 20, no. 4: 23.

180 "The storm knocked down . . . "Storm Damage Survey Hints Park Better-Worse," Tacoma *News Tribune* Sunday Ledger, October 21, 1962, A-11; and author interview with Tacoma parks employee Kathy Allen, May 2013.

180 "Not all the historic . . . " "Missing Senators Stymie Election," *Oregon Statesman*, February 13, 1959, A-18; and author interview with R. Gregory Nokes, Oregon author of *Breaking Chains, Slavery on Trial in the Oregon Territory* (Corvallis: Oregon State University Press, 2013).

180 "Claims adjustors in Oregon and . . . " Lucia, *The Big Blow*, 59–60.

180 "The governor requested $2.7 million . . . " "Hatfield Asks for $2.7 Million in Aid," *Oregonian*, October 16, 1962, A-9.

180 "A preoccupied president . . . " "Agreement for Federal Aid Is Signed Monday," *Oregon Statesman*, October 23, 1962, A-3.

180 "Capell's KGW television station did not . . . " Mabel Capell's October 18 letter to Anton and Laimi Wagner.

EPILOGUE

181 " . . . Capell stayed on the air . . . " Tomlinson, "Predictor of 'Storm of the Century' Dies," *Oregonian* , June, 16, 2009, B-2.

181 "If that had been me . . . " Author interview with Capell coworker Doug LaMear, June 2015.

181 "In the later years, as Capell's . . . " Raymond R. Hatton, *Portland, Oregon Weather and Climate: A Historical Perspective* (Bend, OR: Geographical Books, 2005), 30.

181 "It was a lesson in love . . . " Floyd McKay tribute to Capell posted at the *Oregonian* website, June 15, 2009 (see at connect.oregonlive.com/user/ floydmckay /index.html).

182 "Thunderstorms rumbled through . . . " Stuart Tomlinson, "Jack Capell, The Forecaster Who Predicted the Columbus Day Storm, Dies," June 15, 2009 (oregonlive.com/weather/2009/06/jack_capell).

182 "After he retired, Capell received . . . " Capell obituary in the *Seattle Times*, June 26–27, 2009.

182 "John Capell, a television news producer . . . " Author interview with John Capell, March 2018; video transcript of KGW-TV tribute to Jack Capell, June 28, 2009 (kgw.com).

182 "Female aviation pioneer Ruth Wikander . . . " Author interviews with Wayne Nutsch in April, May, and September 2014; National Transportation Safety Board accident report SEA68A0051, Docket Number 3 2058, filed July 29, 1968 (available online at the NTSB aviation data base ntsb.gov/_layouts/ ntsb.aviation/Month.aspx); and Federal Aviation Administration rule governing instruction in single-control aircraft found in Title 14 Code of Federal Regulations 91.109(a) as amended in 2011.

182 "After one year and one thousand hours . . . " "The Colorful History of Aerocar N103D" (aerocarforsale.com/history.htm).

183 "Taylor, a pilot who worked . . . " "Moult Taylor Aerocar Aircraft History, Performance and Specifications" (pilotfriend.com/aircraft%20performance/ aerocar.htm).

183 "Called by many the cultural . . . " Munk, "Francis J. Murnane: Oregon's Forgotten 'Rabble-Rousing, Art-Loving Longshoreman,'" unpublished ms, 1–4.

183 "Union leaders had urged . . . " Anne Saker, "A Forgotten Labor Leader, Forgotten Wharf," *Oregonian*, April 22, 2009, B-1.

184 "An unofficial review of the weather bureau's . . . " "2 Aviation Weather Forecasters At Bureau Office Write 'Unofficial Autopsy' Of Columbus Day Storm," *Oregonian*, October 10, 1963), H-8.

184 "Holcomb retired from the National Weather Service . . . " Author interview with Jim Holcomb, July 2015.

184 "A case in point is the . . . " "Advance Warning of October 12th Wind Storm,"
       October 19, 1962, memo from John F. Hagemann, emergency planning
       engineer for the Oregon State Highway Department, to Tom Edwards,
       assistant state highway engineer.

185 "The sergeant was probably referring . . . " "Weather Bureau Cites Special
       Wind Warning," *Oregon Statesman*, October 18, 1962, A-12.

185 "I'm not blaming the weather bureau . . . " "Schools Had No Warning of
       Hurricane," *Oregon Statesman*, October 19, 1962, A-7.

185 "A careful review of the events . . . " Capell, *Terrible Tempest*.

186 "If Civil Defense can't come forward . . . " Sam Frear, "Cone Asks: What
       Good Is CD?" Eugene *Register-Guard*, October 17, 1962, A-1.

186 "The Ash Wednesday Storm of March 6–8, 1962 . . . " "The Historical Context
       of Emergency Management" (training.fema.gov/), chapter 1, pages 3–5.

187 "We would have anywhere from . . . " Author interview with Buehner, March
       2013.

187 "Since the 1990s we have predicted . . . " Mass, Columbus Day Storm
       presentation at University of Washington, October 11, 2012.

187 "The UW's Mass sounded . . . " Mass, "Warning: Major Storms Threaten the
       Pacific Northwest," October 11, 2016 (cliffmass.blogspot.com).

187 "The National Weather Service . . . " Mass, "A Deeper Look at Saturday's
       Storm," October 18, 2016 (cliffmass.blogspot.com).

188 "The low-pressure center . . . " Mass, "A Short Comment about the Storm,"
       October 16, 2016 (cliffmass.blogspot.com).

188 "Some of the forecasting uncertainty . . . " Mass, "A Deeper Look," October
       18, 2016 (cliffmass.blogspot.com).

189 "We have more people . . . " Author interview with Buehner, March 2013.

Bibliography

NEWSPAPERS

Bellingham Herald
Capital Journal (Salem)
Corvallis Gazette-Times
Register-Guard (Eugene)
Herald (Everett)
Humboldt Standard (Eureka, CA)
Morning Oregonian
Newberg Graphic
Newport News
New York Times
Oregon Statesman (Salem)
Oregonian
San Francisco Chronicle
Seattle Daily Times
Seattle Post-Intelligencer
Seattle Times
Sunday Olympian (Olympia, WA)
News Tribune (Tacoma)
Tillamook Headlight Herald
Columbian (Vancouver, WA)
Province (Vancouver, BC)
Vancouver Sun (Vancouver, BC)

WEBSITES

aerocarforsale.com
atmos.washington.edu/videos/columbusday
barometer.ws/history
boeing.com/history

bornfreeusa.org

cliffmass.blogspot.com

climate.washington.edu

columbian.com/history

currentresults.com/weather/Oregon/annual-days-of-sunshine.php

digitalcommons@linfield.edu

HistoryLink.org

historynet.com

hydeparkhistory.org

idahoforests.org

legendsofamerica.com

metoc.ndbc.noaa.gov.JWTC

modis.gsfc.nasa.gov

newportoregon.gov

northwest-wine.com

ntsb.gov

oldsmokeys.org

oregon.gov/ODA

oregonencyclopedia.org

OregonLive.com

pdxradio.com

peterearley.com

pilotfriend.com

portlandoregon.gov/parks

princeofpinot.com

nsrh.noaa.gov

radomes.org

salemhistory.net

seattleweekly.com

science1.nasa.gov/missions/tiros/

spaceneedle.com

thenationalpasttimemuseum.com

training.fema.gov

usatoday.com

wrcc.dri.edu

wunderground.com, yagrs.org

BOOKS

Abbott, Carl. *Greater Portland: Urban Life and Landscape in the Pacific Northwest.* Philadelphia: University of Pennsylvania Press, 2001.

Abbott, Carl. *Portland in Three Centuries: The Place and the People.* Corvallis: Oregon State University Press, 2011.

Anderson, Rick. *Seattle Vice: Strippers, Prostitution, Dirty Money and Crooked Cops in the Emerald City.* Seattle: Sasquatch Books, 2010.

Arno, Stephen F., and Ramona P. Hammerly. *Northwest Trees: Identifying and Understanding the Region's Native Trees.* Seattle: Mountaineers Books, 2007.

Becker, Paula, Alan Stein, and the HistoryLink staff. *The Future Remembered: The 1962 World's Fair and Its Legacy.* Seattle: Seattle Center Foundation, 2011.

Berger, Knute. *Space Needle: The Spirit of Seattle.* Seattle: Documentary Media, 2012.

Blalock, Barner. *Portland's Lost Waterfront.* Charleston, SC: History Press, 2012.

Burt, Christopher. *Extreme Weather.* New York: W.W. Norton, 2004.

Capell, Jack. *Surviving the Odds.* Claremont, CA: Regina Books, 2007.

Carson, Rob. *Mount St. Helens: The Eruption and Recovery of a Volcano.* Seattle: Sasquatch Books / Tacoma: News Tribune, 2015.

Chasan, Daniel Jack. *The Water Link: A History of Puget Sound as a Resource.* Seattle: Washington Sea Grant Program, University of Washington, 1981.

Conrad, Joseph. *Typhoon.* 1st Forge ed. New York: Tim Doherty, 1999 [originally published in 1903].

Daniel, John. *The Far Corner: Northwestern Views on Land, life and Literature.* Berkeley: Counterpoint Press, 2009.

DeBlieu, Jan. *Wind: How the Flow of Air Has Shaped Life, Myth, and the Land.* New York: Houghton Mifflin, 1998.

de Villiers, Marq. *Windswept: The Story of Wind and Weather.* New York: Walker, 2007.

Franklin, Dorothy. *West Coast Disaster: Columbus Day 1962.* Portland: Gann Publishing, 1962.

Goeres-Gardner, Diane L. *Inside Oregon State Hospital.* Charleston, SC: History Press, 2013.

Hatton, Raymond R. *Portland, Oregon, Weather and Climate: A Historical Perspective.* Bend: Geographical Books, 2005.

Huler, Scott. *Defining the Wind.* New York: Three Rivers Press, 2004.

Kingle, Matthew. *Emerald City: An Environmental History of Seattle.* New Haven, CT: Yale University Press, 2007.

Laskin, David. *Rains All the Time.* Seattle: Sasquatch Books, 1997.

Lehr, Paul E. *Storms: The Origins and Effects.* New York: Golden Press, 1969.

Lucia, Ellis. *The Big Blow.* Portland: New-Times Publishing, 1963.

Mass, Cliff. *The Weather of the Pacific Northwest.* Seattle: University of Washington Press, 2008.

McEntee, Cindy. *Mo's on the Waterfront.* Wilsonville, OR: BackSeller Books, 2004.

Meares, John. *Voyages Made in the Years of 1788 and 1789 from China to the Northwest Coast of America.* New York: Da Capo Press, 1967 [originally published in 1790].

Miller, George R. *Lewis and Clark's Northwest Journey: Weather Disagreeable.* Portland: Frank Amato Publications, 2004.

Miller, George R. *Pacific Northwest Weather.* Portland: Frank Amato Publications, 2002.

Morgan, Murray. *Century 21: The Story of the Seattle World's Fair.* 50th anniv. ed. Bellingham, WA: Chuckanut Editions, 2012 [originally published by University of Washington Press, 1963].

Morgan, Murray. *Puget's Sound: A Narrative of Early Tacoma and the Southern Sound.* Seattle: University of Washington Press, 1979.

Neely, Wayne. *The Great Bahamas Hurricane of 1866.* Bloomington, IN: iUniverse, 2011.

Nicol, Eric. *Vancouver.* Garden City, NY: Doubleday, 1978.

Parzybok, Tye W. *Weather Extremes of the West.* Missoula, MT: Mountain Press, 2005.

Peattie, Donald C. *A Natural History of Western Trees.* Boston: Houghton Mifflin, 1950.

Perry, Vivian, and John Vincent. *Winemakers of the Willamette Valley, Pioneering Vintners from Oregon's Wine Country.* Charleston, SC: American Palate / History Press, 2013.

Plude, Betty Chapman. *Columbus Day Storm 1962 Memories.* Independence, OR: Self-published, 2011.

Price, Richard L. *Newport, Oregon 1866–1936, Portrait of a Coastal Resort.* Newport, OR: Lincoln County Historical Society, 1975.

Punter, John. *The Vancouver Achievement.* Vancouver: University of British Columbia Press, 2003.

Schlosser, Eric. *Command and Control: Nuclear Weapons, the Damascus Accident and the Illusion of Safety.* New York: Penguin Books, 2013.

Sensel, Joni. *Tradition Through the Trees: Weyerhaeuser's First 100 years.* Seattle: Documentary Book, 1999.

Sokol Blosser, Susan. *At Home in the Vineyard: Cultivating a Winery, an Industry, and a Life.* Berkeley: University of California Press, 2006.

Strever, Bill. *And Soon I Heard a Roaring Wind: A Natural History of Moving Air.* New York: Little, Brown and Co., 2016.

Suzuki, David, and Wayne Grady. *Tree: A Life Story.* Vancouver, BC: Greystone Books, 2004.

Van Pelt, Robert. *Forest Giants of the Pacific Coast.* Seattle: University of Washington Press, 2001.

Walker, Gabrielle. *An Ocean of Air: Why the Wind Blows and Other Mysteries of the Atmosphere.* Boston: Houghton, Mifflin, Harcourt, 2007.

Williams, Jack. *The Weather Book.* New York: Vintage Books, 1997.

Winchester, Simon. *Pacific.* New York: Harper Collins, 2015.

Wohlleben, Peter. *The Hidden Life of Trees.* Vancouver, BC: Greystone Books, 2015.

Wyatt, Steve. *The Bayfront Book.* Waldport, OR: Oldtown Printers, 1999.

Zebrowski, Ernest Jr. *Perils of a Restless Planet: Scientific Perspectives on Natural Disasters.* Cambridge, UK: Cambridge University Press, 1997.

JOURNAL ARTICLES, RESEARCH PAPERS, REPORTS, MANUSCRIPTS, AND NEWSLETTERS

Annual Typhoon Report—1962. San Francisco: US Fleet Weather Central/Joint Typhoon Warning Center, January 28, 1963.

Area of Old-Growth Forests in California, Oregon and Washington. Charles L. Bolsinger and Karen L. Waddell. Research Bulletin 197. Portland: US Forest Service, Pacific Northwest Research Station, December 1993.

"Best of the S.O.B.s." John G. Mitchell. *Audubon*, September 1974.

"A Climatological Perspective of the 1962 Columbus Day Storm." Wolf Read. Presented at the Oregon chapter of the American Meteorological Society meeting, Portland, Oregon, October 13, 2012.

"The Columbus Day 'Big Blow' in Oregon." Fred W. Decker, Owen P. Cramer, and Byron P. Harper. *Weatherwise*, December 1962. https://doi.org/10.1080/00431672.1962.9925132.

"The Columbus Day Storm of 1962." Steve Pierce. *Oregon Chapter of the American Meteorological Society Newsletter*, October 2, 2012.

Columbus Day Storm in Washington, October 12, 1962. Earl L. Phillips. Seattle: US Weather Bureau Office, n.d. Retrieved from Washington State Library, January 2013.

"Detailed Analysis of the 1962 Columbus Day Windstorm in Oregon and Washington." Robert E. Lynott and Owen P. Cramer. *Monthly Weather Review* 94, no. 2 (February 1966).

The Economic Impact of the Wine and Grape Industries on the Oregon Economy. Berkeley: Full Glass Research, January 2015.

"Extreme Waves under Hurricane Ivan." David W. Wang, Douglas A. Mitchell, William J. Teague, Ewa Jarosz, and Mark S. Hulbert. *Science* 309 (August 2005). http://archive.li/bD8pX#selection-635.0-665.15.

"Extreme Weather Trends over the Pacific Northwest." Cliff Mass. Presented at the Northwest Climate Conference, University of Washington, Seattle, September 9, 2014.

"Fifty Years Later: Legacy of Columbus Day Storm Still Stands." Mark Floyd. Corvallis: Oregon State University News and Research,

October 2, 2012. http://today.oregonstate.edu/archives/2012/oct/
fifty-years-later-legacy-columbus-day-storm-still-stands.

Five Year Forest Inventory and Analysis Report: Washington 2002–2006.
Research Bulletin GTR-800, April 2010. Portland: US Forest Service Pacific
Northwest Station.

Five Year Forest Inventory and Analysis Report: Oregon 2001–2005. Research
Bulletin GTR-765, November 2008. Portland: US Forest Service Pacific
Northwest Station.

"Francis J. Murnane: Oregon's Forgotten 'Rabble-Rousing, Art-Loving Longshoreman.'" Michael Munk. Portland: Unpublished manuscript, June 2012.

"Geology of Vineyards in the Willamette Valley, Oregon." George W. Moore.
Corvallis, OR: Oregon State University, Department of Geosciences,
updated September 16, 2002. http://cmug.com/chintimp/Willamette.
vineyards.htm.

"The Great Columbus Day Blowdown." John Clark Hunt. *American Forests*,
January 1963.

"The Guardian Class Radar Picket Ships." Unpublished manuscript. Compiled
by YAGR's Radar Picket Ship Association.

"How Will Climate Change Affect Explosive Cyclones in the Extratropics of
the Northern Hemisphere?" Christian Seiler and F. W. Zwiers. *Climate
Dynamics*, August 12, 2015.

Hull of a Good Story. Nicole Montesano. McMinnville: Oregon Wine
Press, September 1, 2012. http://www.oregonwinepress.com/
article?articleTitle=hull-of-a-good-story-1346360315-1310—food.

"Japanese Wartime Incarceration in Oregon." In *Oregon Encyclopedia of History and Culture*. Portland: Portland State University/Oregon Historical
Society.

*A Life Sentence: The Sad and Dangerous Realities of Exotic Animals in Private
Hands in the U.S.* Sacramento, CA: American Protection Institute, February 23, 2006.

Major Southern California Windstorms (1858–November 2013). 2014 Natural
Hazards Mitigation Plan, City of Newport Beach, California, Section 10,
Table 10-4.

"Monitoring and Understanding Changes in Extremes: Extratropical Storms,
Winds, and Waves," Russell S. Vose, Scott Applequist, et al. *Bulletin of the
American Meteorological Society*, March 2014.

National Transportation Safety Board. Report SEA68A0051, Docket Number
3 2058, September 25, 1968.

New Horizons. Newsletter of the women's aviation group, Ninety-Nines, Inc.,
November-December 2008.

Official United States Weather Bureau. Weather advisory issued by Portland
weather bureau office at 10:10 a.m., October 12, 1962.

Official United States Weather Bureau. Wind warning issued by Portland
weather bureau office at 10:40 a.m., October 12, 1962.

OldSmokeys Newsletter. Portland: Pacific Northwest Forest Service Association, Fall 2012.

Oregon Agripedia. Salem, OR: Oregon Department of Agriculture annual report, 2015.

Oregon State Hospital Admission Report. January 1962–June 1962. Submitted by hospital superintendent Dean Brooks to the Oregon State Board of Control.

Oregon State Hospital Monthly Report. October 1962. Submitted by hospital superintendent Dean Brooks to the Oregon State Board of Control, November 1, 1962.

PinotFile. Online newsletter compiled by William "Rusty" Gaffney. Archived at princeofpinot.com.

"The Rise and Fall of the Pacific Northwest Log Export Market." Jean M. Daniels. General Technical Report 624. Portland: United States Forest Service, Pacific Northwest Research Station, February, 2005.

Summary of the 1962 Columbus Day Storm in Oregon. Gilbert Sternes. Portland: US Weather Bureau Office, November 23, 1962.

"Tales from the Hoh and Quileute." Albert Reagan and L. V. W. Walters. *Journal of American Folklore* 46, no. 182 (1933). doi:10.2307/535636.

The Terrible Tempest of the Twelfth. Jack Capell. Portland: Pioneer Broadcasting Company, 1962.

The Third Oregon Climate Assessment Report. Corvallis: Oregon State University, Oregon Climate Change Research Institute, January 2017.

Timber Harvest Report 2014. Olympia: Washington State Department of Natural Resources, August 2015.

Timber Harvest Report 2014. Salem: Oregon State Department of Forestry, July 2015.

Western Conservation Journal (Special Blowdown Issue) 20 (4). Seattle: Evergreen Publications, 1963.

Windthrown Timber Survey in the Pacific Northwest 1962. P. W. Orr. Portland: Pacific Northwest Region, United States Forest Service, March 1963.

Index

Note: Photographs and maps are indicated by italicized page numbers. References to the Columbus Day Storm are indicated by the use of "Storm" capitalization. Index entries for Source Note material are indicated by the page number followed by an "n."